ADVANCES IN BIOMOLECULAR MEDICINE

PROCEEDINGS OF THE 4TH BIBMC (BANDUNG INTERNATIONAL BIOMOLECULAR MEDICINE CONFERENCE) 2016 AND THE 2ND ACMM (ASEAN CONGRESS ON MEDICAL BIOTECHNOLOGY AND MOLECULAR BIOSCIENCES), BANDUNG, WEST JAVA, INDONESIA, 4–6 OCTOBER 2016

Advances in Biomolecular Medicine

Editors

Robert Hofstra
Departement of Clinical Genetics, Erasmus MC University Medical Center, Rotterdam, The Netherlands

Noriyuki Koibuchi
Department of Integrative Physiology, Gunma University Graduate School of Medicine, Gunma, Japan

Suthat Fucharoen
Thalassemia Research Centre, Institute of Molecular Biosciences, Mahidol University, Thailand

CRC Press is an imprint of the
Taylor & Francis Group, an **informa** business

A BALKEMA BOOK

CRC Press/Balkema is an imprint of the Taylor & Francis Group, an informa business

© 2017 Taylor & Francis Group, London, UK

Typeset by V Publishing Solutions Pvt Ltd., Chennai, India
Printed and bound in Great Britain by CPI Group (UK) Ltd, Croydon, CR0 4YY

All rights reserved. No part of this publication or the information contained herein may be reproduced, stored in a retrieval system, or transmitted in any form or by any means, electronic, mechanical, by photocopying, recording or otherwise, without written prior permission from the publisher.

Although all care is taken to ensure integrity and the quality of this publication and the information herein, no responsibility is assumed by the publishers nor the author for any damage to the property or persons as a result of operation or use of this publication and/or the information contained herein.

Published by: CRC Press/Balkema
　　　　　　　P.O. Box 11320, 2301 EH Leiden, The Netherlands
　　　　　　　e-mail: Pub.NL@taylorandfrancis.com
　　　　　　　www.crcpress.com – www.taylorandfrancis.com

ISBN: 978-1-138-63177-9 (Hbk + CD-ROM)
ISBN: 978-1-315-20861-9 (eBook)

Advances in Biomolecular Medicine – Hofstra, Koibuchi & Fucharoen (Eds)
© 2017 Taylor & Francis Group, London, ISBN 978-1-138-63177-9

Table of contents

Preface — vii

Iron overload intolerance in Balb/c mice — 1
N. Anggraeni, M.R.A.A. Syamsunarno, R.D. Triatin, D.A. Setiawati, A.B. Rakhimullah, C.C. Dian Irianti, S. Robianto, F.A. Damara, D. Dhianawaty & R. Panigoro

Mapping of health care facilities in the universal coverage era at Bandung District, Indonesia — 5
N. Arisanti, E.P. Setiawati, I.F.D. Arya & R. Panigoro

Influence of phosphatidylcholine on the activity of SHMT in hypercholesterolemic rats — 9
A. Dahlan, H. Heryaman, J.B. Dewanto, F.A. Damara, F. Harianja & N. Sutadipura

Molecular detection of DHA-1 AmpC beta-lactamase gene in *Enterobacteriaceae* clinical isolates in Indonesia — 13
B. Diela, S. Sudigdoadi, A.I. Cahyadi, B.A.P. Wilopo, I.M.W. Dewi & C.B. Kartasasmita

Correlation between natrium iodide symporter and c-fos expressions in breast cancer cell lines — 19
A. Elliyanti, N. Noormartany, J.S. Masjhur, Y. Sribudiani, A.M. Maskoen & T.H. Achmad

Fat mass profile in early adolescence: Influence of nutritional parameters and rs9939609 FTO polymorphism — 23
S.N. Fatimah, A. Purba, K. Roesmil, G.I. Nugraha & A.M. Maskoen

A study of the palatal rugae pattern as a bioindicator for forensic identification among Sundanese and Malaysian Tamils — 29
R. Khaerunnisa, M. Darjan & I.S. Hardjadinata

Effect of acrylamide in steeping robusta coffee (*Coffea canephora* var. *robusta*) on memory function and histopathological changes of brain cells in male rats (*Rattus norvegicus*) — 33
D.Y. Lestari, M. Bahrudin, Rahayu, Fadhil & A.W. A'ini

Effects of aerobic exercise and a high-carbohydrate diet on RBP4 expression in rat skeletal muscle — 37
N. Najmi, Y. Sribudiani, B.S. Hernowo, H. Goenawan, Setiawan, V.M. Tarawan & R. Lesmana

Correlation of physical activity and energy balance with physical fitness among the professors of the University of Padjadjaran — 41
J. Ninda, L. Lubis, A. Purba, I.B. Akbar, W. Karhiwikarta, Setiawan, V.M. Tarawan, R. Farenia, G.I. Nugraha, M. Rizky Akbar, D.K. Sunjaya, S. Rachmayati, I. Ruslina, Hanna, T. Hidayat, P. Tessa, N. Sylviana, Ronny, T. Nurhayati, Y.S. Pratiwi, N.V. Utami, Juliati, S.N. Fatimah, Y. Indah & F. Huda

Molecular biology of irreversible pulpitis: A case report — 47
D. Prisinda & A. Muryani

Effect of different doses of X-ray irradiation on survival of human esophageal cells — 53
I.M. Puspitasari, R. Abdulah, M.R.A.A. Syamsunarno & H. Koyama

The effect of mesenchymal stem cells on the endothelial cells of diabetic mice — 57
A. Putra, A. Rahmalita, Y. Tarra, D.H. Prihananti, S.H. Hutama & N.A.C. Sa'diyah

Effects of selenium on SePP and Apo B-100 Gene expressions in human primary hepatocytes — 61
M. Putri, N. Sutadipura, S. Achmad, C. Yamazaki, S. Kameo, H. Koyama, M.R.A.A. Syamsunarno, T. Iso & M. Kurabayashi

Inhibition of cAMP synthesis abolishes the impact of curcumin administration in the skeletal muscle of rodents *H.R.D. Ray & K. Masuda*	65
Effect of Monosodium Glutamate (MSG) on spatial memory in rats (*Rattus norvegicus*) *R. Razali, S. Redjeki & A.A. Jusuf*	71
Mycobacterium tuberculosis load and rifampicin concentration as risk factors of sputum conversion failure *E. Rohmawaty, H.S. Sastramihardja, R. Ruslami & M.N. Shahib*	75
Interleukin-22 serum in comedonal acne vulgaris: Proof of inflammation *K. Ruchiatan, R. Hindritiani, E. Sutedja & S. Maulinda*	79
Pharmacokinetic optimization of the treatment of TB meningitis with TB drugs *R. Ruslami*	83
Iron-chelating effect of *Caesalpinia sappan* extract under conditions of iron overload *R. Safitri, D. Malini, A.M. Maskoen, L. Reniarti, M.R.A.A. Syamsunarno & R. Panigoro*	87
The role of *S.aureus* and *L.plantarum* as an immunomodulator of IFNα macrophages and fibronectin dermal fibroblast secretion *R.S.P. Saktiadi, S. Sudigdoadi, T.H. Madjid, E. Sutedja, R.D. Juansah, T.P. Wikayani, N. Qomarilla & T.Y. Siswanti*	93
Exon globin mutation of β-thalassemia in Indonesian ethnic groups: A bioinformatics approach *N.I. Sumantri, D. Setiawan & A. Sazali*	99
Serum immunoglobulin-E level correlates with the severity of atopic dermatitis *O. Suwarsa, E. Avriyanti & H. Gunawan*	105
Fatty liver in fasted FABP4/5 null mice is not followed by liver function deterioration *M.R.A.A. Syamsunarno, M. Ghozali, G.I. Nugraha, R. Panigoro, T. Iso, M. Putri & M. Kurabayashi*	109
The potential of seluang fish (*Rasbora* spp.) to prevent stunting: The effect on the bone growth of *Rattus norvegicus* *Triawanti, A. Yunanto & D.D. Sanyoto*	113
Effect of cryoprotectants on sperm vitrification *R. Widyastuti, R. Lesmana, A. Boediono & S.H. Sumarsono*	119
Mucoprotective effect of *Trigona* propolis against hemorrhagic lesions induced by ethanol 99.5% in the rat's stomach *V. Yunivita & C.D. Nagarajan*	123
Author index	127

Preface

Following our 1st BIBMC (Bandung International Biomolecular Conference) in 2010, 2nd BIBMC in 2012 and 3rd BIBMC in 2014, the 4th BIBMC is organized in conjunction with the 2nd ACMM (ASEAN Congress on Medical Biotechnology & Molecular Biosciences) following the 1st ACMM in Bangkok in 2015. We are grateful to Mahidol University, University of Philippines & University Science Malaysia for supporting Padjadjaran University to hold this joint scientific meeting in Bandung, West Java, Indonesia. In line with our new regional commitment to the SDGs (Sustainable Development Goals) the theme of 4th BIBMC & 2nd ACMM is 'Medical innovation & translational research to ensure healthy lives & promote well-being for all at all ages'.

Preceeding this joint scientific meeting on 4th-6th October 2016 at Trans Luxury Hotel, we conducted laboratory workshops at our Faculty of Medicine, Padjadjaran University Campus on (1) Thalassemia Diagnostic (2) Flowcytometry & ELISA (3) HE & Immunochemistry Staining (4) 3-D Cell Culture (5) RNA Isolation, RT PCR, DNA Isolation. The joint meeting covers various sessions on Infection, Oncology, Tuberculosis, Genetic, Thalassemia, Nutrition, Cardiovascular, Wound Healing & Endocrinology. The sessions consist of 30 invited speakers covering various countries e.g. Malaysia, Singapore, Philippines, Thailand, Japan, Netherlands, Germany, Colombia, Indonesia. Additional free sessions are comprised of 16 oral and 76 poster presentations representing 31 domestic & overseas higher education & research institutions. Our national vaccine institute, Bio Farma hosted a luncheon seminar on Vaccine Currrent Update.

We are very grateful to distinguished international speakers: Dr Badrul Hisham Yahya, Prof Bladimiro Rincon-Orozco, Dr Brijesh Kumar Singh, Prof Carmencita Padilla, Dr Chong Pei Nee, Prof Dean Nizetic, Dr Edward Rob Aarnoutse, Prof Harry Surapranata, Dr Jin Zhou, Dr Marc Hubner, Prof Noriyuki Koibuchi, Dr Oraphan Sripichai, Prof Paul Yen, Prof Prasert Auwewarakul, Dr Rohit Sinha, Prof Suthat Fucharoen, Prof Toshiharu Iwasakai & Prof Zilfalil Alwi for their state of the art lectures.

This 4th BIBMC & 2nd ACMM Joint Meeting wouldn't be possible without generous supports from Prof Akmal Taher (Staff to Indonesian Minister of Health), Prof Tri Hanggono Achmad (Rector of Padjadjaran University), Prof Sangkot Marzuki (Chairman, Indonesian Academy of Sciences), Prof Amin Subandrio (Director of Eijkman Institute), Dr Yoni Fuadah (Dean of Faculty of Medicine, Padjadjaran University) & IDI (Indonesian Medical Association).

Special appreciation goes to CRC Press/Balkema of the Taylor & Francis Group for publishing 'ADVANCES IN BIOMOLECULAR MEDICINE, Proceedings of 4th BIBMC & 2nd ACMM' Submitted for indexing to Scopus & Thomson Reuters with generous recommendations by Prof Noriyuki Koibuchi (Gunma University), Prof Robert Hofstra (Erasmus MC University) & Prof Suthat Fucharoen (Mahidol University).

We are grateful to our internal reviewers Dr Bachti Alisjahbana, Dr Rovina Ruslami, Dr Edhyana Sahiratmadja, Dr Andri Rezano, Dr Afiat Berbudi, Dr Fathul Huda, Dr Astrid Feinisa Khaerani, Dr Hasan Bashari & Dr Alvinsyah Adhityo Pramono for their hard work in the selection of the papers to be presented at 4h BIBMC & 2nd ACMM.

We are also very grateful to our generous sponsors: Biofarma, Medco Group, Santosa Hospital, Prodia Clinic Laboratory, The Trans Group, Sentra Diagnostik Dinamika, Batik Komar, Hasan Sadikin Hospital & National Eye Centre—Cicendo Eye Hospital.

We hope that through 4th BIBMC & 2nd ACMM we keep improving our capacities in advancing biomolecular medicine, medical biotechnology & molecular biosciences in ASEAN Countries and beyond as part of our global commitment to the sustainable development goals.

January 2017
4th BIBMC & 2nd ACMM Committee
Prof Ramdan Panigoro, Dr Rizky AA Syamsunarno,
Dr Nur Atik, Dr Ronny Lesmana

Iron overload intolerance in Balb/c mice

N. Anggraeni
Department of Biochemistry and Molecular Biology, Faculty of Medicine, Universitas Padjadjaran, Bandung, West Java, Indonesia
Medical Laboratorium Technologyst of Bakti Asih School of Analyst, Bandung, West Java, Indonesia

M.R.A.A. Syamsunarno, R.D. Triatin, D.A. Setiawati, A.B. Rakhimullah, C.C. Dian Irianti, S. Robianto, F.A. Damara, D. Dhianawaty & R. Panigoro
Department of Biochemistry and Molecular Biology, Faculty of Medicine, Universitas Padjadjaran, Bandung, West Java, Indonesia

ABSTRACT: The iron–diabetes hypothesis has been supported by a number of epidemiological studies showing that body iron deposits are positively associated with blood glucose levels. The underlying pathological interactions between iron and glucose homeostasis are not well understood. A recent study used C57BL/6 mice that were injected with a high dose of iron to establish an iron overload mouse model. However, in Indonesia, most of the experiments involving mouse models use the Balb/c strain instead of C57BL/6. The objective of this study was to validate the effects of iron overload in Balb/c mice. Mice were intraperitoneally injected with iron dextran (10 mg/μL per day, iron overload group). Fasting blood glucose level, body weight, and liver weight were measured after 7 days of treatment. After 7 days, only six out of 12 mice survived. The body weights of the iron overload group were 18.5% lower than that of the control group, with liver enlargement and lower blood glucose levels. In conclusion, our study indicated that Balb/c mice are intolerant to excessive amounts of iron.

1 INTRODUCTION

Repetitive blood transfusions in patients with a chronic disease such as thalassemia will inevitably lead to iron overload that cannot be actively removed from the body (Eliezer A. Rachmilewitz & Patricia J. Giardina, 2011). Iron overload is the major cause of morbidity in thalassemia patients. Even non-transfused patients develop iron overload secondary to increased intestinal absorption of dietary iron. Iron overload is a leading cause of mortality and organ injury (Borgna-Pignatti C & Gamberini MR, 2011).

The heart, along with the liver and the endocrine glands, is the main organ affected by excess iron accumulation, and thus iron-loading conditions are primarily manifested as cardiac dysfunction and failure, liver dysfunction and cirrhosis, and endocrine abnormalities including hypothyroidism, hypogonadism, and diabetes mellitus, as well (Galaris, D. & Pantopoulos, K. 2008).

Iron deposits in thalassemia patients, who have received multiple blood transfusions, can exceed the storage and detoxification capacity of ferritin. Moreover, the excess iron produces fully saturated transferrin. Consequently, "free" iron (or non-transferrin bound iron, NTBI) begins to accumulate in tissues and blood. This "free" iron can catalyze the formation of very toxic compounds, such as the hydroxyl radical (–OH), from compounds such as hydrogen peroxide, which are normal metabolic by products (Fenton reaction) (Andrews, N.C, 2005). The resulting oxidative stress is associated with damage of cellular macromolecules, tissue injury, and disease (Wang, Jian. & Pantopoulus, Kostas, 2011, Moon, Se Na. et al, 2011).

Epidemiological studies have shown that body iron stores are positively correlated with serum insulin and blood glucose levels, which supports the iron–diabetes hypothesis (Jia, Xuming. et al, 2013). The high prevalence of type 2 diabetes observed in individuals with hereditary hemochromatosis (HH) further supports the idea that systemic iron loading is associated with impaired glucose metabolism (Gujja, P. et al, 2010, Hatunic M, et al.). However, the underlying pathological interactions between iron and glucose homeostasis are not well understood.

A recent study used C57BL/6 mice that were injected with a high dose of iron to establish an iron overload mouse model (G. Ramey et al. 2007, Moon, Se Na. et al. 2011). However, in Indonesia, most of the experiments involving mouse models use the Balb/c strain instead of C57BL/6. The objective of this study was to validate the effects of iron overload in Balb/c mice to better understand the relationship between iron metabolism and glucose homeostasis.

2 MATERIALS AND METHODS

2.1 Animals

Male Balb/c mice (Animal Laboratorium, Department of Pharmacology and Therapy, Faculty of Medicine, Universitas Padjadjaran, Bandung, Indonesia) were fed *ad libitum* with a standard chow diet. Mice were divided into two experimental groups. The first group was intraperitoneally injected with saline (100 μL per day, control group; MERCK, USA) for 7 days, and the second group was intraperitoneally injected with iron-dextran (10 mg/μL per day, iron overload group; SANBE, Indonesia) for 7 days from 8 weeks of age.

2.2 Measurement of body weight

Body weight was measured with the Nagata Scale (Japan) every 3 days, starting from the first day of the study.

2.3 Measurement of fasting blood glucose levels

After 3 days of injection, blood was collected after 6 hours of fasting for glucose determination. Blood glucose level was measured through the tail tip using the GlucoDr Super Sensor AGM 2200 (Allmedicus, Japan). Fasting blood glucose level was measured every 3 days for 1 week.

2.4 Measurement of liver weight

After 7 days, mice were killed by cervical dislocation, and liver weight was measured using the Nagata Scale (Japan).

3 RESULTS

3.1 Intolerance of Balb/c mice to iron overload

After 7 days of injection, six out of the 12 mice died in the iron overload group. This shows that the mortality of Balb/c mice intolerant to iron overload is 50%. We suggest that intolerance of Balb/c mice to iron overload is caused by hypoglycemia and organ damage.

3.2 Weight loss of iron overload group

Body weight of mice was measured every 3 days, starting from the first day of the study. The iron overload group had lower body weight after 3 days of injection.

The average body weight of the iron overload group was 18.5% lower than that of the control group (Figure 1). This difference can be attributed to the increased metabolic rate (Andrews, N.C, 2005).

3.3 Iron overload group had a hypoglycemic condition

The average fasting glucose concentration of the iron overload group was 104.7 md/dL, which was lower than that of the control group. Our study, together with the data reported in the literature, suggests that, for some unknown reasons, mice are resistant to the toxic effects of iron-induced organ damage (Hentze, M. W., et al, 2010).

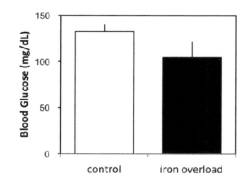

Figure 2. The average fasting glucose concentration of the control group and the iron overload group.

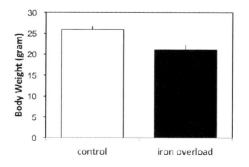

Figure 1. The average body weight of the control group and the iron overload group.

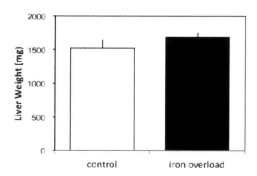

Figure 3. Liver weight of the iron overload group and the control group.

3.4 Liver enlargement in the iron overload group

Our study showed that liver weight in the iron overload group was larger (9.5%, hypertrophy) than the control group (Figure 3). The average liver weight of the iron overload group was 1.69 g and that of the control group was 1.53 g.

4 DISCUSSION

Iron is an essential nutrient that plays a role in important cellular and tissue functions including oxygen transport, nucleotide synthesis, mitochondrial respiration, and host defense. It is absorbed from the diet by duodenal enterocytes that release iron into the plasma through ferroportin, the only known cellular iron exporter (Brissot P et al, 2012). However, iron is also potentially toxic, because, under aerobic conditions, it catalyzes the production of ROS (reactive oxygen species) and the generation of highly reactive radicals (e.g. the hydroxyl radical) through Fenton chemistry (Koppenol, W. H, 1993).

Iron is a biologically essential element, acting as a cofactor for many enzymes involved in the oxidation–reduction process and other types of metabolism (Oudit GY, 2006). In our study, it was observed that the body weight of the iron overload group was lower than that of the control group, which we presume is caused by the increased metabolic rate (Andrews, N.C, 2005).

In the state of iron overload, non-transferrin-bound iron moves in and out of the cells by diffusion and generates free iron, forming reactive oxygen species that result in cell or tissue damage (Crichton RR. et al, 2002). Particularly, it damages the liver and heart, resulting in complications such as liver failure and cardiomyopathy, of which cardiac failure is most fatal (Mamtani M & Kulkarni H, 2008).

The control of iron homeostasis by hepcidin represents a classical endocrine regulatory system. Hepcidin regulates iron, and in turn, hepcidin production is regulated by iron in the circulation and in liver stores (Ramos E. et al, 2011): when iron is abundant, hepcidin production is increased to limit dietary iron absorption and release from the stores; when iron is required, a decrease in hepcidin production allows iron to enter the plasma to meet the iron demand (Kautz L. et al, 2014).

Our study shows liver enlargement in the iron overload group, which may represent a compensatory mechanism of iron overload.

Typically, iron loading is associated with insulin resistance. Some studies have reported insulin resistance as a primary consequence of iron overload, while others have suggested that both insulin deficiency and insulin resistance contribute to glucose intolerance (Kim CH, et al, 2011, J. Arezes & E. Nemeth, 2015). In contrast, our study shows that Balb/c mice injected with a high dose of iron dextran had a hypoglycemic condition (Figure 2). The study suggests that, for some unknown reasons, mice are resistant to the toxic effects of iron-induced diabetes.

5 CONCLUSION

Our study indicated that Balb/c mice are intolerant to excessive amounts of iron, which may be associated with liver damage and hypoglycemic condition.

REFERENCES

Aisen, P., Enns, C. and Wessling-Resnick, M. 2001. Chemistry and biology of eukaryotic iron metabolism. Int. J. Biochem. Cell Biol. 33, 940–959.

Andrews, N.C. 2005. Molecular control of iron metabolism. Best Pract. Res. Clin. Haematol. 18, 159–169.

Arezes, J. and E. Nemeth. 2015. Hepcidin and iron disorders: new biology and clinical approaches. Int. Jnl. Lab. Hem. 37, 92–98.

Arredondo M, Fuentes M, Jorquera D, Candia V, Carrasco E, Leiva E, Mujica V, Hertrampf E, Perez F. 2011. Cross-talk between body iron stores and diabetes: iron stores are associated with activity and microsatellite polymorphism of the heme oxygenase and type 2 diabetes. Biol Trace Element Res. 143, 625–636.

Awai M, Narasaki M, Yamanoi Y, Seno S. 1979. Induction of diabetes in animals by parenteral administration of ferric nitrilotriacetate. Am J Pathol. 95. 663–674.

Borgna-Pignatti C, Gamberini MR. 2011. Complications of thalassemia major and their treatment. Expert Rev Hematol. 4, 353–366.

Brissot P, Ropert M, Le LC, Loreal O. 2012. Non-transferrin bound iron: a key role in iron overload and iron toxicity. Biochim Biophys Acta. 1820, 403–410.

Chern JP, Lin KH, Lu MY, Lin DT, Lin KS, Chen JD, Fu CC. 2001. Abnormal glucose tolerance in transfusion-dependent beta-thalassemic patients. Diabetes Care. 24, 850–854.

Crichton RR, Wilmet S, Legssyer R, Ward RJ. 2002. Molecular and cellular mechanisms of iron homeostasis and toxicity in mammalian cells. J Inorg Biochem. 91, 9–18.

Eliezer A. Rachmilewitz and Patricia J. Giardina. 2011. How I treat thalassemia. Blood. 118, 3479–3488.

Galaris, D. and Pantopoulos, K. 2008. Oxidative stress and iron homeostasis: mechanistic and health aspects. Crit. Rev. Clin. Lab. Sci. 45, 1–23.

Grunblatt E, Bartl J, Riederer P. 2011. The link between iron, metabolic syndrome, and Alzheimer's disease. J Neural Trans. 118, 371–379.

Gujja, P. et al. 2010. Iron Overload Cardiomyopathy, Better Understanding of an Increasing Disorder. J Am Coll Cardiol. 56, 1001–1012.

Hatunic M, et al. Effect of iron overload on glucose metabolism in patients with hereditary hemochromatosis. Metabolism. 59, 380–384.

Hentze, M.W., Muckenthaler, M.U., Galy, B. and Camaschella, C. 2010. Two to tango: regulation of mammalian iron metabolism. Cell 142, 24–38.

Jia, Xuming. et al. 2013. Glucose metabolism in the Belgrade rat, a model of iron-loading anemia Am J Physiol Gastrointest Liver Physiol. 304, G1095–G1102.

Kautz L, Jung G, Valore EV, Rivella S, Nemeth E, Ganz T. 2014. Identification of erythroferrone as an erythroid regulator of iron metabolism. Nat Genet. 46, 678–684.

Kim CH, Kim HK, Bae SJ, Park JY, Lee KU. 2011. Association of elevated serumferritin concentration with insulin resistance and impaired glucose metabolism in Korean men and women. Metab Clin Exp. 60, 414–420.

Koppenol, W.H. 1993. The centennial of the Fenton reaction. Free Radical Biol. Med. 15, 645–665.

Liu P, Olivieri N. 1994. Iron overload cardiomyopathies: new insights into an old disease. Cardiovasc Drugs Ther. 8, 101–110.

Mamtani M, Kulkarni H. 2008. Influence of iron chelators on myocardial iron and cardiac function in transfusion-dependent thalassaemia: a systematic review and meta-analysis. Br J Haematol. 141, 882–890.

Moon, Se Na. et al. 2011. Establishment of Secondary Iron Overloaded Mouse Model: Evaluation of Cardiac Function and Analysis According to Iron Concentration Pediatr Cardiol. 32, 947–952.

Niederau C, et al. 1985. Survival and causes of death in cirrhotic and in noncirrhotic patients with primary hemochromatosis. New Engl J Med. 313, 1256–1262.

Oudit GY, Trivieri MG, Khaper N, Liu PP, Backx PH. 2006. Role of L-type Ca^{2+} channels in iron transport and iron-overload cardiomyopathy. J Mol Med. 84, 349–364.

Ramey, G. et al. 2007. Iron overload in Hepc1 –/– mice is not impairing glucose homeostasis FEBS Letters. 58, 1053–1057.

Ramos E, Kautz L, Rodriguez R, Hansen M, Gabayan V, Ginzburg Y, Roth MP, Nemeth E, Ganz T. 2011. Evidence for distinct pathways of hepcidin regulation by acute and chronic iron loading in mice. Hepatology. 53, 1333–1341.

Wang, Jian. and Pantopoulus, Kostas. 2011. Review article: Regulation of cellular iron metabolism. Biochem. J. 434, 365–381.

… …

Mapping of health care facilities in the universal coverage era at Bandung District, Indonesia

N. Arisanti, E.P. Setiawati & I.F.D. Arya
Department of Public Health, Faculty of Medicine, Universitas Padjadjaran, West Java, Indonesia

R. Panigoro
Department of Biochemistry, Faculty of Medicine, Universitas Padjadjaran, West Java, Indonesia

ABSTRACT: The implementation of Universal Health Coverage should be accompanied by increasing and distributing health resources such as human resources, infrastruture, etc in primary health center and hospitals. In this UHC era, Government of Bandung District has made various efforts to strengthen health systems in order of fulfillment the number of health personals. The purpose of this study was to identify health care facilities according to the number, status, type, accreditation, partnership with National Health Insurance (BPJS) and map the distribution.

The method was descriptive study. The data of registered health care facilities was collected from District Health Office and Professional Organization, categorized according to the number, type, location, status, accreditation, and the mapping was performed using the software. The research was conducted in Bandung District, from October to December 2015.

There are 1027 health care facilities in 31 sub-districts. A total of 59.5% is private health care facilities and majority is midwives (28.7%). There are only 5 health care facilities accredited and all are general hospitals. Health care facilities that have formed a new partnership with BPJS reached 29.8%. From 31 sub-districts, Rancaekek is the sub-districts with the highest number of health care facilities.

Keywords: distribution, health care facilities, mapping

1 INTRODUCTION

The access to health care facilities is a necessary condition to achieve Universal Health Coverage. UHC aims to meet population needs for quality health care, remove financial barriers to health care access, reduce incidence of catastrophic health expenditures, attain national and internationally agreed health goals, and ultimately contribute to poverty alleviation and development. (Sambo et al. 2014) Many efforts in the implementation of the UHC to meet the goal should be accompanied by increasing and equitable distribution of resources such as human resources for health, health care facilities and infrastructure, affordability of drugs, medical supplies in health. Equal and proper distribution of health and human services is an effort that can support the achievement of UHC and help in maximizing accessibility, thereby helping government and stakeholders save cost on providing infrastructure for the entire population and most importantly optimize delivery of health care goals. (Ayuba et al. 2016)

As one of the largest archipelagos in the world, Indonesia facing problem on geographical access to health care and unevenly distributed health care facilities. Doctor could not reach the whole population in the working area and visa versa. Their inequitable distribution over space is of concern and has brought about the issue of provision and effective utilization of the facilities. The regular monitoring of access to health care facilities is often a weak component of country and global monitoring the performance of health system. Annual reviews of health sector progress and performance at national and subnational levels, based on a broad set of indicators that cover all areas of performance, should include up-to-date and accurate information on service delivery. There is continuing discussion in Indonesia about the need for improved information on resources for health at the district level where programs are actually delivered. Research by Heywood, et al in 2009 describes the distribution of health care facilities and personnel in the 15 Districts in Java in 2007. The result of this study was about half of the third group of professionals (doctors, nurses, midwives) is a Public Servant. The private sector as the main work is very important for doctors (37%) and more important for midwives (10%). For those who work in the government sector, two-thirds of the doctors and nurses working in health care facilities, while most midwives working at the village level facilities. (Heywood et al. 2009) With UHC, government of Indonesia plans to cover the entire population in 2019. It was widely anticipated that there would be a significant increase in demand for health services in line with the implementation of this program. Although UHC is a program of the central government, its implementation occurred in districts/cities and at the level of fulfillment of the facility through the efforts of the health system in the public and private sectors. Bandung District as one of the districts in West Java has made various efforts to strengthen health systems through improving infrastructure and fulfill of health personnel in various government health care facilities and private sector. In order to improve fulfill the health care facilities in district, it is necessary to identify and map the distribution of health care facilities. Mapping enables professionals to understand complex spatial relationships visually so as to plan effectively and efficiently. Previous study estimated that nearly 80% of the information need of local health system decision and policy makers involves geographical positioning. (Heywood et al. 2009) The purpose of this study was to identify health care facilities according to the number, status, type, accreditation, partnership with National Health Insurance (BPJS) and map the distribution.

2 METHOD

This was a descriptive study to analyze data regarding the number, status, type, accreditation, partnership with BPJS and distribution of health care facilities in Bandung District. The type of health care facilities included in the study was primary and secondary care both public and private as well as military/police and state firms, doctor, dentist and midwife's practice. The data of registered health care facilities was collected from District Health Office and Professional Organization.

The data was analyzed and presented in frequency distribution table. The map of health care facilities was generated by identifying and locating their locations using the software. The study was conducted in October to December 2015 in Bandung District.

3 RESULTS

3.1 Health care facilities

The study covers 31 sub-districts in Bandung District with total area coverage of 1.767,93 Km² and population of 3.470.393. From the data obtained, there are 1027 health care facilities. Most of health care facilities (59,5%) are private sectors (Table 1).

Based on Figure 1, private midwife is the largest group in the primary level. To support the successful implementation of UHC private midwife should be encourage to collaborate with primary physician as a gate keeper.

There are only 5 health care facilities accredited and all are general hospitals, while all primary health cares have not been accredited. Health care facilities that have formed a new partnership with BPJS reached 29.8%.

3.2 Mapping of health care facilities

From the 31 sub-districts, Rancaekek is a sub-district with the highest number of health care facilities, followed by Cileunyi and Baleendah (Figure 2). When compared to population in each

Table 1. Distribution of health care facilities based on owner status.

Type of health care facilities	Number	Percentage
Government	406	39,5
Private	611	59,5
Military	8	0,8
Provincial state	1	0,8
Other	1	0,1
Total	1027	100

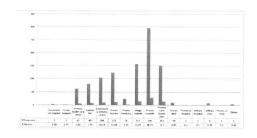

Figure 1. Distribution of health care facilities based on type of health care.

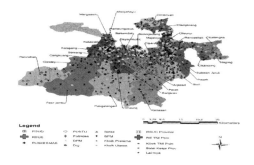

Figure 2. Mapping of health care facilities based on sub-district in Bandung District.

6

sub-districts, it is correlated with the number of population in those sub-districts. Provider density (number of doctors, nurses and midwives/1000 population) was low by international standards.

4 DISCUSSION

Reliable and timely information about resources for health is important to envisioning a health system that can respond to the health challenges. Such information will provide the data indicated that the health system and the health needs of the population are changing and that government must modify policies in response to these changes. In Bandung District, most of health care facilities are private sectors. There is an increasing number of private sectors who do not work for the government at all, and the government has little information about them as well. There is great concern about the lack of attention to human resources in the health sector globally, especially that many governments do not have even basic information about their most important resource: how many health professionals, their age and sex, or how they are distributed. (Heywood et al. 2011) Private practice as the most health care facilities and health personals is now quite important, especially for doctors, and increasingly so for midwives. Previous studies indicated a growing private sector, working either for private facilities (private hospitals, clinics) or in their own private practice without an appointment with the government. (Heywood et al. 2009)

Based on type, most of health care facilities are private midwives. Previous study in Indonesia stated that solo practice of outpatient services comprises 86% of all the health facilities—they are all private. At the sub district level there is a public health center, the satellite health centers, and a much larger number of solo-provider outpatient facilities through which doctors, nurses and midwives operate their afterhours private practices. Village midwives located in many villages also operate private practices. (Heywood et al. 2009) Thus, the private practices became an important source of healthcare facilities.

Based on the Table 2, there are only 5 health care facilities accredited and all are general hospitals, while all primary health cares have not been accredited. This indicates that the health service in primary health care is not standardized yet. Bandung District Health Office had to work hard to be able to make the all health care facilities in the region accredited in 2019. With more than 1000 health care facilities and length of the average of 250 health care facilities must be accredited per year achieved in 4 year. Health care facilities that have formed a new partnership with BPJS reached

Table 2. Distribution of health care facilities based on accreditation status and partnership with BPJS.

	Accreditation status		Partnership with BPJS	
	Number	Percentage	Number	Percentage
Unknown	1	0,1	157	15,3
Yes	5	0,5	306	29,8
No	1021	99,4	564	54,9
Total	1027	100	1027	100

29.8%. In general that have not established partnerships with BPJS is primary health care. BPJS needs to encourage primary health care as a partner in health care delivery.

Access to health facility is critical to health service and distribution as well to meet goal of UHC. (Ayuba et al. 2016) It was agreed that to achieve better access, health care facilities needed to be distributed among the people and the facilities needed to be adequately staffed. (Kumar et al. 2013)

This study identified the available of health care facilities in the study area using the secondary data and presented on map of Bandung District indicating health care facilities' positions. Mapping enables professionals to understand complex spatial relationships visually so as to plan effectively and efficiently. (Abbas et al. 2014 & Bukhari et al. 2013) The evenly distribution of health care facility will be effective and efficient in delivery health service to the people. From the map of Bandung District, it shows that the higher population in sub districts the higher density of health care facilities. Those areas are close to the capital of West Java Province.

As conclusion, in Universal Health Coverage era, the government should anticipate the increasing demand for services significantly in line with the implementation of this program. Bandung District Health Office in collaboration with BPJS needs to encourage private health care facilities to collaborate and cooperate with BPJS in order to achieve the universal coverage.

REFERENCES

Abbas, Idowu Innocent, Abdulqadir, Hussain Zaguru & Bello, Mohammed Nanoh. 2014. Mapping the Spatial Distribution of Health Care Facilities of the Millennium Development Goals (MDGs) in Kaduna North and South Local Governments, Kaduna State, Nigeria. *Global Journal of Human-Social Science: B.* 14(5):1–7.

Ayuba. IGU, Wash. PM. 2016. Identification and Mapping of Health Facilities in Bukuru Town, Plateau State Nigeria. *Journal of Environment and Earth Science.* 6(3):81–94.

Bukhari. AS, Muhammed. I. 2013. Geospatial Mapping of Health Facilities in Yola, Nigeria. (IOSR-JESTFT). 7(3):79–85.

Heywood P, Harahap N, Aryani S. 2011. Recent changes in human resources for health and health facilities at the district level in Indonesia: evidence from 3 districts in Java. *Human Resources for Health*. 9(5):1–6.

Heywood, P., & Harahap, N. 2009. Health facilities at the district level in Indonesia. *Australia and New Zealand Policy*. 6(13):1–11.

Heywood, P., & Harahap, N. 2009. Human resources for health at the district level in Indonesia: the smoke and mirrors of decentralization. *Human Resources for Health*. 7(6):1–16.

Kumar P, Khan AM. 2013. Human Resources Management in Primary Care. *Health and Population: Perspectives and Issues*. 36 (1 & 2):66–76.

Sambo, LG. Kirigia M. 2014. Investing in health systems for universal health coverage in Africa. *BMC International Health and Human Rights*. 14(28):1–22.

Influence of phosphatidylcholine on the activity of SHMT in hypercholesterolemic rats

A. Dahlan, H. Heryaman, J.B. Dewanto, F.A. Damara & F. Harianja
Department of Biochemistry and Molecular Biology, Faculty of Medicine Universitas Padjadjaran, Bandung, West Java, Indonesia

N. Sutadipura
Department of Biochemistry and Molecular Biology, Faculty of Medicine Universitas Padjadjaran, Bandung, West Java, Indonesia
Department of Biochemistry, Universitas Islam Bandung, Bandung, Jawa Barat, Indonesia

ABSTRACT: Hypercholesterolemia is one of the risk factors of cardiovascular diseases. To elucidate the correlation between hypercholesterolemia and deficiency of phosphatidylcholine, animal experimental study was conducted. Wistar rats aged 8 weeks were randomly distributed into three groups: control group (standardized normal diet), high-fat diet group, and phosphatidylcholine-supplemented high-fat diet group. Total plasma cholesterol and SHMT activity were measured in the liver of rats by using spectrophotometry.

After 8 weeks, total plasma cholesterol was 72.77 mg% and SHMT activity was 68.93 units/mg in the control group. In the high-fat diet group, total plasma cholesterol was 94.72 mg% and SHMT activity was 150.07 units/mg. After supplementation of phosphatidylcholine, total plasma cholesterol was 78.47 mg% and SHMT activity was 80.97 units/mg. From these results, it can be concluded that phosphatidylcholine supplementation can protect rats from high-fat diet induced-hypercholesterolemia.

Keywords: Hypercholesterolemia, phosphatidylcholine, Serine hydroxymethyltransferase enzyme

1 INTRODUCTION

Dyslipidemia is a condition in which there is an increase in cholesterol and triacylglycerol levels that exceed the normal range, and it is one of the risk factors of atherosclerosis (Assman, 1982; Assman 1993; Askandar, 1994, Nelson et al., 2000). A study conducted in Framingham (Gordon et al., 1977) and Helsinki (Manninen et al., 1992) has proved that the risk of atherosclerosis is closely related to high- and low-density lipoprotein levels. Low-Density Lipoprotein (LDL) is positively correlated, whereas High-Density Lipoprotein (HDL) is negatively correlated with the incidence of atherosclerosis (Nelson et al., 2000).

Lipoprotein is made up of lipids (triacylglycerol, cholesterol, and phospholipid) and certain proteins such as apoprotein (Assman, 1993; Harper, 1995; Nelson et al., 2000). HDL is composed of 3–6% triacylglycerol, 14–18% cholesterol ester, 45–50% protein, and 20–30% phospholipid. The highest fraction of phospholipid in HDL is phosphatidylcholine (74.6%) (Assman, 1983). Plasma HDL plays an important role in cholesterol homeostasis both in the cell and plasma. Its function is to transport cholesterol back to the liver from the periphery, fibroblast, and arterial smooth muscle to be later broken down into bile acid and used in the digestive tract (Assman, 1993, Nelson et al, 2000).

In the cholesterol "reverse transport" process, phosphatidylcholine (lecithin) is required as a cofactor and substrate of Lecithin Cholesterol Acyl Transferase (LCAT). Phosphatidylcholine also acts as the activator of Cholesterol Ester Hydrolase and as the inhibitor of Acyl-CoA Cholesterolacyltransferase (ACAT) (Howard and Patelski, 1976). Both enzymes are important for intracellular cholesterol metabolism. There is a positive correlation of LCAT and ACAT with plasma lipids (Patelski et al., 1976). An increase in LCAT activity during plasma lipid metabolic disturbance has shown the importance of phosphatidylcholine.

In the plasma HDL of a healthy individual, the phosphatidylcholine and sphingomyelin ratio is 4:1 (Peeter, 1976). Furthermore, it has been established that in individuals with hypercholesterolemia, this ratio changes with an increase in phosphatidylcholine and a decrease in sphingomyelin. This leads to the need for increasing phosphatidylcholine in patients with hypercholesterolemia.

In mammals, there is a close correlation between glycine metabolism and biosynthesis of phospholipids. Glycine is made from choline or serine (Harper, 1995, Voet and voet, 1995). Serine hydroxymethyltransferase (SHMT) is an enzyme (Lehninger, 1985; Harper, 1995) that is required for the interconversion of glycine and choline.

This enzyme needs pyridoxal phosphate (PLP) and tetrahydrofolate (FH4) as coenzymes. FH4 helps catalyze the reaction of glycine–serine interconversion and the formation of the C1 unit (single carbon unit) in the form of N5,N10-metilen FH4. This reaction is the main source of C1 in mammals. C1 units are very important in the biosynthesis of various elements, one of which is choline. In a study conducted by Sutadipura (1996), it has been found that SHMT activity is increased in hypercholesterolemia induced by a high-fat diet. This shows that under such a condition, glycine–choline metabolism and biosynthesis of phospholipids are increased.

The purpose of this study is to determine the influence of phosphatidylcholine on hypercholesterolemia. To address this problem, phosphatidylcholine was administered to Wistar rats with or without high-fat diet-induced hypercholesterolemia.

2 MATERIALS AND METHODS

2.1 Rats and sample collection

Wistar rats aged 4 to 5 weeks old were used in this study. Rats were housed in a temperature-controlled environment under a 12-hour light/12-hour dark cycle, and given free access to water and standard chow. The animals were divided into three groups: control group, High-Fat Diet (HFD) group and phosphatidylcholine-supplemented HFD group. The treatment was started after 7 days of acclimatization. The control group was given standard chow and the HFD group was given a mixture of 1% cholesterol, 1% palmitic acid, 5% yolk, 1% butter, and standard chow up to 100%. Phosphatidylcholine supplementation was given 90 mg/100 g of body weight. After 8 weeks of treatment, the rats were euthanized and blood from the heart and liver was collected.

2.2 Measurements of plasma cholesterol and SHMT activity

Plasma was collected from blood, and total plasma cholesterol was measured by the Liebermann–Burchard procedure as described previously (Wood et al., 1980). SHMT activity was measured from the liver as described previously (Omura dan Sato, 1964).

2.3 Statistical analysis

One-way ANOVA was performed for three samples, and Bonferroni's *post hoc* multiple comparison tests were performed to evaluate the difference between the control and experimental groups. A p value of less than 0.05 was considered as statistically significant.

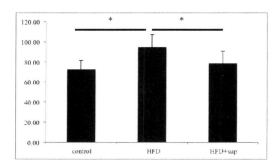

Figure 1. Plasma cholesterol after 8 weeks of experiment. *p < 0.05. HFD = high-fat diet; sup = phosphatidylcholine supplementation.

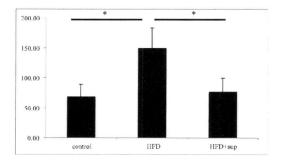

Figure 2. SHMT activity after 8 weeks of experiment. *p < 0.05. HFD = high-fat diet; sup = phosphatidylcholine supplementation.

3 RESULTS

After 8 weeks, total plasma cholesterol in the control, HFD, and phosphatidylcholine-supplemented HFD groups were 72.77 mg%, 94.72 mg%, and 78.47 mg%, respectively (Figure 1). SHMT activity was two times higher in the HFD group, but the activity reduced to 50% after phosphatidylcholine supplementation (Figure 2).

4 DISCUSSION

High-fat diet can cause high blood cholesterol levels. Plasma cholesterol, mainly LDL, is taken up by tissues. When intracellular cholesterol levels exceed the normal range, it can trigger: 1. inhibition of *de novo* cholesterol synthesis by the inhibition of Hydroxymethylglutaryl-CoA reductase (HMG CoA-reductase), 2. activation of ACAT that plays a role in the intracellular esterification process of cholesterol, 3. inhibition of LDL receptor synthesis at the RNA level.

The aforementioned three events are the main cause of the variant responses of individual experimental animals to a high-fat diet. The reduction of total plasma cholesterol in rats fed with a high-fat diet after phosphatidylcholine supplementation can be explained by: 1. the increase in the activity of intracellular cholesterol ester hydrolase that hydrolyzes the storage of cholesterol ester into free cholesterol that effluxes from the cell (Howard and Patelski, 1976), 2. the inhibition of ACAT that plays a role in the esterification of cholesterol, 3. the activation of LCAT that plays a role in cholesterol reverse transport (Assman, 1976; Harper, 1999).

LCAT plays a role in the esterification of HDL-cholesterol, and phosphatidylcholine acts as the source of fatty acids in the process. Cholesterol ester is transported to the liver to be metabolized into bile acid. So, the phosphatidylcholine acts on LCAT, ACAT, and cholesterol ester hydrolase to control the plasma cholesterol level.

The reduction of total plasma cholesterol in rats with HFD-induced hypercholesterolemia by supplementation of phosphatidylcholine shows that there is a deficiency of phosphatidylcholine in the HFD group. One of the cholesterol synthesis parameters is SHMT activity.

We showed that under the conditions of HFD-induced hypercholesterolemia, SHMT activity was increased in the liver of rats. This indicated the correlation between hyperlipidemia and increasing SHMT activity, probably by increased glycine–choline metabolism that resulted in excess choline.

Phosphatidylcholine supplementation in rats fed with a HFD showed reduced cholesterol levels and SHMT activity. This clearly showed that there is a deficiency of phosphatidylcholine under the conditions of HFD-induced hypercholesterolemia.

5 CONCLUSION

Phosphatidylcholine supplementation can protect rats from hypercholesterolemia and reduce SHMT activity after high-fat diet-induced-hypercholesterolemia.

REFERENCES

Askandar, T. (1994). Dislipidemia-Diabetes melitus-Aterosklerosis. *In Full Script of Simposium Nasional Diabetes dan Lipid,* Surabaya: 177.

Assman, G. (1982). Lipid metabolism and atherosclerosis. *Central Laboratory of the Medical Faculty University of Munster Federal Rep of Germany*, 1–5, 16–18, 28–53, 76–100.

Assman, G. (1993). Lipid metabolism disorders and coronary heart disease. *Central Laboratory of the Medical Faculty University of Munster Federal Rep of Germany*, 24–29, 32–33, 71–72, 97–99, 211–216.

Bender, D.A. (1985). Amino acid metabolism 2nd ed. *Chicester-NewYork John Wiley & Sons*, 47, 80, 95–100.

Breslow, L. (1996). New mouse models of lipoprotein disorder and atherosclerosis International edition. *Science,* 272(5262):685–8.

Brewer, H.B. Jr. et al. (1996). Genetic dyslipoproteinemias International Edition.

Fuster, V et al. Atherosclerosis and caronary artery disease. *Lippincort-Raven Publishers*, Philadelphia: 363–375.

Fuster, V. et al. Atherscrosclerosis and coronary artery disease. *Lippincort-Raven Publishers*, Philadelphia: 69–83.

Ginsberg, H.N. (1994). Lipoprotein Metabolism And The Relationship To Atherosclerosis In: Hunninghake DB, Ed. Lipid Disorder. *Clinics of North America*, North America: 78 no. 1:1.

Gordon, T., Castelli, W.P., Hortland, M.S., Kannel W.B. (1997). High Density Lipoprotein As A Protective Factor Against Coronary Heart Disease, The Farmingham Study. *Am. J. Med.* 62(5), 707–14.

Harper Ft. A. (1995). Review Of Physiological Chemistry. 18 th ed. *Lange Maruzen*, Tokyo: 112–125, 348–356.

Howard, AN. and Patelski, J. (1976). Effect Of EPL On The Lipid Metabolism Of The Arterial And Other Tissues. In peeters ed. Phosphatidylcholine; Biochemical And Clinical Aspect Of Essential Phospholipids, *Berlin-Heidelberg-New-York, Springer-Verlag*, 187–200.

Lehninger, A.L., Nelson D.L., and Cox, M.M. (1996). Principles of biochemistry. 2nd Ed. *New York Worth publisher*, 674–678.

Lekim, D. (1974). On The Pharmacokinetics of Orally Applied Essentiale Phospholipids. In Peeters H. ed. Phosphatidilcholine: Biochemical And Aspect Of Essentiale Phospholipids. *Berlin-Heidelberg- New York; Springer Verlag*, 48–63.

Lusis J. Aldons et al. (1998). Genetics of Atherosclerosis. In Topol J. Eric ed. Texbook of Cardiovascular Medicine. *Lippincort-Raven Publishers*, Philadelphia: 2389–2408.

Mahley, R.W. (1978). Alterations In Plasma Lipoprotein Induced By Cholesterol Feeding In Animals Including Man. In: Dietscky JM, Ed, Disturbances in lipid and Lipoprotein Metabolism. *Bethesda Md*, 181.

Manninem, V., Teskanen, L., Konkinen P. (1992). Joint Effect Of Serum Trigliseride, LDL cholesterol and HDL Cholesterol Consentralion On Coronary Heart Disease Risk In The Helsinky Heart Study: Implication For Treatment. *Circulation*, 85: 37–45.

Nelson, D. L. & Cox, M.M., (2000). Lehninger Principles of Biochemistry, 3rd ed. *Worth Publishers*, 768–778, 804–811.

Nugraha, S. (1986). Komposisi Fraksi Fosfolipid Lipoprotein Dengan Densitas Tinggi (HDL) dari Plasma Penderita Hiperlipoproteinemia. *Tests Magister Sains Fakvltas Pascasarjana UNPAD Bandung*, Bandung: 87–88.

Nugraha, S. (1996). Aktifitas Serin Hidroksimetil Trans/ Erase Pada Tikus Yang Menderita Hiperlipoproteinemia. *Disertasi Program Pascasarjana Universitas Padjadjaran Bandung*, Bandung: 43–51,88–90.

Omura, T & Sato, R. (1964) The Carbon Monoxide-Binding Pigment Of Liver Microsomes. I. Evidence For Its Hemoprotein Nature. *J Biol Chem,* 239:2370–8.

Peeters, H. (1974). The Biological Significance Of The Plasma Phospholipids. In: Peeters H. Ed. Phosphatidilcholine: Biochemical And Clinical Aspect Of Essential phospholipids. *Berlin Heidelberg -New York Springer Verlag,* 12–90.

Samochowiec, L. (1974). On the Action of Essential Phospholipids in Experimental Atherosclerosis. In: Peeters H. ed. Phosphatidylcholine. Biochemical and Clinical Aspect of Essential Phospholipids. *Berlin - Heidelberg - New york Springer Verlag,* 211–227.

Schaefer, E.J.(1990). High Density Lipoprotein and Coronary Heart Disease. *Gower Med Publisher,* New-York: 2–3, 18–19.

Steel, R.G.D.dan Torrie J.H. (1989). Prinsip dan Prosedur Statistika. *PT. Gramedia,* Jakarta: 168–178, 228–232.

Stryer, Rubert. (1995). Biochemistry, 4th ed. *W.H. Freeman and Company*, New York: 631,633–634, 685–700, 719–723.

Suryasubrata, S. (1983). Metodologi Penelitian. Jakarta: 48–50, 79–80.

Voet, D. and Voet J.G. (1995). Biochemitry. *New-York-Chichester-Brisbane Wiley & Sons*, 316–318, 662–689, 699–702, 713–715, 736–763.

Wills Eric D. (1989). Wills' Biochemical basis of Medicine 2nd ed. *Bristol Wright,* 146–148, 229, 237, 245.

Wood, P.D., Bachorik, P.S., Albers, J.J., Stewart, C. C., Winn, C & Lippel, K. (1980) Effects of sample aging on total cholesterol values determined by the automated ferric chloride-sulfuric acid and Liebermann-Burchard procedures. *Clin Chem,* 26, 592–7.

Molecular detection of DHA-1 AmpC beta-lactamase gene in *Enterobacteriaceae* clinical isolates in Indonesia

B. Diela
Faculty of Medicine, Universitas Padjadjaran, Bandung, Indonesia

S. Sudigdoadi, A.I. Cahyadi, B.A.P. Wilopo & I.M.W. Dewi
Department of Microbiology and Parasitology, Faculty of Medicine, Universitas Padjadjaran, Bandung, Indonesia

C.B. Kartasasmita
Department of Pediatrics, Faculty of Medicine, Universitas Padjadjaran, Bandung, Indonesia
Dr. Hasan Sadikin Hospital, Bandung, Indonesia

ABSTRACT: Dhahran (DHA)-1 is an inducible plasmid-mediated AmpC beta-lactamase gene, which was reported as the most frequent cause of escalating morbidity and mortality rate among patients with infectious diseases. The aim of this study was to determine the prevalence of bla_{DHA-1} gene in *Enterobacteriaceae* clinical isolates in Indonesia. A total of 85 *Enterobacteriaceae* clinical isolates, including *E. coli, Klebsiella. pneumoniae, E. cloacae, K. oxytoca, E. aerogenes*, and *Proteus mirabilis*, were included in this study. The samples were collected from several private hospitals and clinical laboratories in Bandung, Indonesia, during a period between July 2014 and April 2015. The molecular detection of bla_{DHA-1} gene was carried out by PCR assay. We found that 32.9% of *Enterobacteriaceae* clinical isolates carried bla_{DHA-1} gene, which was predominantly found in *K. pneumoniae* (75%) followed by *K. oxytoca* (50%), *E. cloacae* (28.6%), and *E.coli* (14%). These results revealed a higher frequency of bla_{DHA-1} gene detected in *Enterobacteriaceae* clinical isolates in Indonesia compared with those in surveillance studies conducted worldwide. To our knowledge, this is the first molecular epidemiology report from Indonesia that showed the high prevalence of bla_{DHA-1} gene in *Enterobacteriaceae* clinical isolates.

Keywords: AmpC Beta-Lactamase; DHA-1; *Enterobacteriaceae*; PCR; Indonesia

1 INTRODUCTION

Inadequate regulation of antibiotic therapy for many infectious diseases in most Asian countries, including Indonesia, may increase resistance spectrum of gram-negative bacteria to beta-lactam antibiotics.[1] With the increasing use of beta-lactam antibiotics and various inhibitor combinations, such as amoxicillin-clavulanic acid or sulbactam, production of AmpC beta-lactamase enzyme by gram-negative bacteria have emerged.[2] This enzyme can hydrolyze and inactivate the third generation of beta-lactam antibiotics, including oxyimino-cephalosporins (cefotaxime, ceftazidime, and ceftriaxone), monobactam (aztreonam), and cephamycins (cefoxitin and cefotetan).[3] AmpC beta-lactamase enzyme has been mainly detected in *Enterobacteriaceae* family of gram-negative bacteria and frequently causes severe infection cases in hospitalized patients.[4] This enzyme is poorly inhibited by many common beta-lactamase inhibitors, such as clavulanic acid and sulbactam, but can only be inhibited by AmpC inhibitors, such as cloxacillin or 3-aminophenylboronic acid.[2,3]

Detection of AmpC beta-lactamase enzyme is important to improve the clinical treatment for patients with infection due to *Enterobacteriaceae*. Recently, several phenotypic tests for detection of this enzyme are being developed with varying degrees of success, such as AmpC Disk Test, 3-Aminophenylboronic Acid Disk Test, and Three-Dimensional Test.[5,6] However, there is no specific phenotypic test to confirm the presence of AmpC beta-lactamases in clinical isolates which has been validated by Clinical Laboratory Standards Institute (CLSI).[7] Therefore, there is a need to conduct an examination using PCR (Polymerase Chain Reaction) as the gold standard method to detect the presence of AmpC beta-lactamase gene in *Enterobacteriaceae* clinical isolates.[2,7]

The AmpC beta-lactamase gene (bla_{AmpC}) consists of 6 families: ACC, FOX, MOX, DHA, CIT and EBC. These are encoded either in chromosome or plasmid.[3,8] The expression of bla_{AmpC} chromosomal gene can be triggered by the presence of beta-lactam antibiotics, especially cefoxitin.[3,7] The plasmid-mediated bla_{AmpC} gene is derived from the bla_{AmpC} chromosomal gene or can be transferred horizontally between plasmids of *Enterobacteriaceae*.[3,5,9] DHA-1 is a subfamily of DHA, an inducible plasmid-mediated bla_{AmpC} gene, which has been reported as the most frequent cause of escalating morbidity and mortality rate among infected patients compared with other families of plasmid-mediated bla_{AmpC} gene.[10,12] Bla_{DHA-1} gene was originated from the chromosomal bla_{AmpC} gene of *Morganella morganii*, which the expression can be induced by the presence of beta-lactam antibiotics. Thus, the increasing use of beta-lactam antibiotics in many of infection cases may induce the expression of bla_{DHA-1} gene among *Enterobacteriaceae* clinical isolates.[3,10,11]

At present, there has not been any molecular epidemiological data in Indonesia that report the prevalence of bla_{DHA-1} gene in *Enterobacteriaceae*. Therefore, this study was designed to detect the presence of bla_{DHA-1} gene in *Enterobacteriaceae* clinical isolates in Indonesia by using PCR assay. The results will make health care providers to be more careful in administering antimicrobial therapy for infected patients to decrease the occurrence of antimicrobial resistance and the expansion of resistance spectrum of *Enterobacteriaceae* to beta-lactam antibiotics.

2 MATERIALS AND METHODS

2.1 Bacterial isolates

Ninety two *Enterobacteriaceae* clinical isolates were collected from several private hospitals and clinical laboratories in Bandung, Indonesia, during a period between Juli 2014 and April 2015.

These isolates were taken from patient's specimens, including blood, urine, abdominal pus, vaginal swab, etc. The organisms were grown on nutrient agar plates and biochemical tests were carried out to identify the bacterial species of the *Enterobacteriaceae* clinical isolates. Eighty five *Enterobacteriaceae* clinical isolates were included in this study, including *Escherichia coli* (n = 50), *Klebsiella pneumoniae* (n = 24), *E. cloacae* (n = 7), *K. oxytoca* (n = 2), *E. aerogenes* (n = 1), and *Proteus mirabilis* (n = 1). Seven unknown or non-specific isolates were, such as *Klebsiella sp.*, *Serratia sp.*, or rod-shaped gram-negative bacteria, were excluded from the study.

2.2 Plasmid DNA isolation

Plasmid DNA from *Enterobacteriaceae* clinical isolates were isolated using the Xprep Plasmid DNA Mini Kit (PhileKorea Technology Inc., Seoul, Korea) according to the manufacturer's instructions.

2.3 PCR amplification of bla_{DHA-1} gene

The oligonucleotide primers used for PCR assay were DHAM-forward (F) (5'-AAC TTT CAC AGG TGT GCT GGG T-3') and DHAM-reverse (R) (5'-CCG TAC GCA TAC TGG CTT TGC-3'), which were designed by Pérez-Pérez FJ[8] and synthesized by Integrated DNA Technology, Inc., Lowa, USA. Reactions were carried out in a MasterCycler Gradient™ (Eppendorf Inc., Hamburg, Germany) in 12.5 μL of mixtures containing 6.25 μL of DreamTaq Green PCR Master Mix™ (Thermo Fisher Scientific Inc., MA, USA), which consists of DreamTaq DNA polymerase, 2X DreamTaq Green Buffer, dNTPs, and 4 mM $MgCl_2$; 0.125 μL of 10 μM DHAM-F Primer; 0.125 μL of 10 μM DHAM-R Primer; 5.5 μL of nuclease-free water; and 0.5 μL of isolated plasmid DNA.

The PCR condition used in this study was; 1 cycle of initial denaturation at 95°C for 2 minutes, followed by 30 cycles of 95°C for 30 seconds, 58°C for 1 minute and t72°C for 1 minute and then a final extension at 72°C for 10 minutes.

2.4 Electrophoresis

12 μL of PCR products were resolved on 1% agarose gels, which was made by 3 mg of TopVision™ Agarose Gel (Thermo Fisher Scientific Inc., MA, USA). For negative control, 12 μL of PCR product without isolated plasmid DNA was added to the well. The ladder used containing 8 μL of nuclease-free water, 2 μL of loading dye, and 2 μL of GeneRuler™ 1kb molecular size DNA Ladder (Thermo Fisher Scientific Inc., MA, USA). The electrophoresis process runs at 80 V for 30 minutes using PowerPac™ Basic Power Supply (Bio-Rad Laboratories, Inc., USA). The gel was then visualized with Ultraviolet Trans illuminator (UVDI, Major Science Inc., USA). The positive result of bla_{DHA-1} gene detection was shown by presence of 405 bp band in electrophoresis gel (Figure 1).

2.5 Data analysis

Data were statistically described in terms of frequencies (number of cases) and relative frequencies (percentages) of bla_{DHA-1} gene detected among *Enterobacteriaceae* clinical isolates. All statistical analysis was performed using computer program Microsoft Excel 2010 (Microsoft Cooperation, NY., USA).

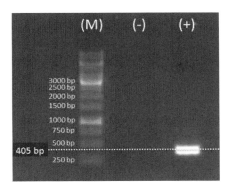

Figure 1. PCR assay for inducible plasmid-mediated bla_{DHA-1} gene (405 bp); (M): GeneRuler™ 1kb molecular size DNA ladder, (−): negative control, (+): positive clinical isolates.

2.6 Ethical considerations

This study was reviewed and approved by the review boards of The Health Research Ethics Committee Faculty of Medicine Universitas Padjadjaran, Bandung, Indonesia, on 23th September 2014, as a part of the study entitled "Detection of Genes Encoding Extended Spectrum Beta-Lactamase in *Enterobacteriaceae* Isolates by Using Loop-Mediated Isothermal Amplification (LAMP)". The project identification code is 533/UN6.C2.1.2/KEPK/PN/2014.

3 RESULTS AND DISCUSSION

Generally, the expression of plasmid-mediated AmpC beta-lactamase gene cannot be induced by the presence of beta-lactam antibiotics; however, DHA-1 is exceptional.[10,11] The expression of bla_{DHA-1} gene in the bacterial plasmid was regulated by AmpR gene, which was linked to its inducible characteristic, same with chromosomally-mediated AmpC beta-lactamase gene.[3,11] The increasing use of beta-lactam antibiotics in many of infection cases may lead to an overexpression of bla_{DHA-1} gene in the plasmid of *Enterobacteriaceae* which can be horizontally transferred to other bacterial plasmids. This mechanism was correlated to the occurrence of antimicrobial resistance in *Enterobacteriaceae* clinical isolates which may cause treatment failure and increase of morbidity and mortality rate of infected patients.[3,10-12] Thus, the detection of bla_{DHA-1} gene in the plasmid of *Enterobacteriaceae* clinical isolates is important to improve the clinical treatment of the infected patients.

In this study, PCR was used as a gold standard method to detect the presence of bla_{DHA-1} gene in *Enterobacteriaceae* clinical isolates.[2,7] The amplicon size of PCR product for this gene was 405 base pairs.[8] As shown in Table 1, PCR amplification revealed a prevalence of 32.9% for bla_{DHA-1} gene encoded among the total eighty five *Enterobacteriaceae* clinical isolates tested in this study. Of these, bla_{DHA-1} gene was predominantly found in *K. pneumoniae* (75%) followed by *K. oxytoca* (50%), *E. cloacae* (28.6%), and *E. coli* (14%).

Table 1. Prevalence of bla_{DHA-1} gene among *Enterobacteriaceae* clinical isolates.

Bacterial species	Total number of clinical isolates	Number of isolates with blaDHA-1 gene (+)	Percentage of isolates with blaDHA-1 gene (+)
E. coli	50	7	14%
K. pneumoniae	24	18	75%
E. cloacae	7	2	28.6%
K. oxytoca	2	1	50%
E. aerogenes	1	0	0%
P. mirabilis	1	0	0%
Total	85	28	32.9%

To our knowledge, this is the first report from Indonesia about the prevalence of bla_{DHA-1} gene among *Enterobacteriaceae* clinical isolates. The prevalence found in this study were higher than those in molecular epidemiological reports published worldwide. In Korea, bla_{DHA-1} gene have been detected in 17.7% of *E. coli* and *K. pneumoniae* clinical isolates,[13] which similarly with the results of study in India (17.8%).[14] In Singapore, 21% of *E. coli* and *K. pneumoniae* clinical isolates have been reported carrying bla_{DHA-1} gene in the plasmids.[15] Compared to those reports, the studies from most of non-Asian countries showed the low prevalence of bla_{DHA-1} gene in *Enterobacteriaceae* clinical isolates. From the surveillance studies in United Kingdom, United States, and Egypt, bla_{DHA-1} gene have been reported in 2.22%, 6.97%, and 6.67% of *Enterobacteriaceae* clinical isolates, respectively.[16-18]

Among *Enterobacteriaceae* clinical isolates tested in in this study, *K. pneumoniae* (75%) was the most prevalent clinical isolates carrying bla_{DHA-1} gene in the plasmid, which is consistent with most of surveillance studies worldwide. However, the frequency of the bla_{DHA-1} gene detected in *K. pneumoniae* clinical isolates found in this study was the highest number compared with most of previous studies.[13-18] The high prevalence of bla_{DHA-1} gene in *Enterobacteriaceae* clinical isolates in Indonesia was thought as a consequence of irrational use of beta-lactam antibiotics in many of infection cases which may induced the expression and spreading of inducible plasmid-mediated bla_{DHA-1} gene from one infected patient to others.

4 CONCLUSIONS

The presence of bla_{DHA-1} gene in the bacterial plasmid has an important mechanism of treatment failure among patients with *Enterobacteriaceae* infection cases.[12] This study was the first report of the high prevalence of bla_{DHA-1} gene among *Enterobacteriaceae* clinical isolates in Indonesia. The inducible plasmid-mediated bla_{DHA-1} gene was detected in 32.9% of *Enterobacteriaceae* clinical isolates, which was predominantly found in *K. pneumoniae* clinical isolates. The results found in this study should increase awareness of health care provider in administering the rational and appropriate antibiotic to their patients. Alternatively, when the presence of inducible plasmid-mediated bla_{DHA-1} gene is detected in the plasmid of infection causal agent, the physician should avoid the use of the third generation of beta-lactam antibiotics and prescribe the fourth generation of beta-lactam antibiotics and/or carbapenems. Therefore, the occurrence of antimicrobial resistance and the expansion of resistance spectrum to beta-lactam antibiotics among *Enterobacteriaceae* could be minimized.

However, this study had certain limitations. 1) This study did not use a positive control of organism carrying bla_{DHA-1} gene for the electrophoresis process; 2) The information about the sources of each clinical isolate used in this study is limited; 3) A relatively small number of *Enterobacteriaceae* clinical isolates included in this study may limit the power of our statistical analysis; 4) The samples used in this study were taken from patients in Bandung local area, which could not represent the prevalence of bla_{DHA-1} gene among *Enterobacteriaceae* clinical isolates in all Indonesian areas. Thus, further studies from other areas in Indonesia may be required to provide additional molecular epidemiology reports about the presence of bla_{DHA-1} gene in *Enterobacteriaceae* clinical isolates.

ACKNOWLEDGEMENTS

This study was supported by grant from Internal Research Grant of Universitas Padjadjaran.

The authors would like to thank staff of Molecular Microbiology Laboratory and Microbiology Laboratory—Faculty of Medicine Universitas Padjadjaran, Bandung, Indonesia for continuously sub-culturing the stored *Enterobacteriaceae* clinical isolates; Gita Widya Pradini, MD, M.Kes., for giving us consultation about molecular detection of bacterial plasmid DNA by using examination of PCR.

AUTHOR CONTRIBUTIONS

B.D and I.M.W.D conceived the research idea and research design. B.D performed the experiments, interpreted the data, and wrote the manuscript. S.S provided and managed the *Enterobacteriaceae* Clinical Isolates for this study. C.B.K and A.I.C was responsible for the critical revision and the final approval of article. B.A.P.W contributed in the preliminary study for PCR optimization. All authors have read and approved the final revision of the manuscript.

CONFLICT OF INTEREST

The authors declare no conflict of interest. The funding sponsors had no role in the design of the study; in the collection, analyses, or interpretation of data; in the writing of the manuscript, and in the decision to publish the results.

REFERENCES

[1] Saharman, Y.R.; Lestari, D.C. Phenotype Characterization of Beta-Lactamase Producing *Enterobacteriaceae* in the Intensive Care Unit (ICU) of Cipto Mangunkusumo Hospital in 2011. *Acta Medica Indonesiana—The Indonesian Journal of Internal Medicine*. 2013, 45, 11–16.

[2] Barua, T; Shariff, M; Thukral, S.S. Detection and Characterization of AmpC B-Lactamases in Indian Clinical Isolates of *Escherichia coli*, *Klebsiella pneumoniae* and *Klebsiella oxytoca*. *Universal Journal of Microbiology Research* 2013, 1, 15–21.

[3] Jacoby, G.A. AmpC β-Lactamases. *Clinical Microbiology Reviews*. 2009, 22, 161–182.

[4] Yusuf, I., Haruna, M. Detection of AMPC and ESBL Producers among *Enterobacteriaceae* in a Tertiary Health Care in, Kano-Nigeria. *International Journal of Science and Technology*. 2013, 3, 220–225.

[5] Doi, Y; Paterson, D.L. Detection of Plasmid-Mediated Class β-Lactamases. *International Journal of Infectious Diseases*. 2007, 11, 191–197.

[6] Peter-Getzlaff, S., Polsfuss, S., Poledica, M., Hombach, M., Giger, J., Böttger, E.C., Zbiden, R., Bloemberg, G.V. Detection of AmpC Beta-Lactamase in *Escherichia coli*: Comparison of Three Phenotypic Confirmation Assays and Genetic Analysis. *Journal of Clinical Microbiology*. 2011, 49, 2924–2932.

[7] Clinical and Laboratory Standards Institute. Performance Standards for Antimicrobial Disk Susceptibility Tests; Approved Standard—Eleventh Edition. *CLSI document M02-A11*. 2011, 32, 25–26.

[8] Pérez-Pérez, F.J., Hanson, N.D. Detection of Plasmid-Mediated AmpC β-Lactamase Genes in Clinical Isolates by Using Multiplex PCR. *Journal of Clinical Microbiology*. **2002**, *40*, 213–2162.

[9] Fam, N., Gamal, D., Said, M.E., Defrawy, I.E., Dadei, E.E., Attar, S.E., Sorur, A., Ahmed, S., Klena, J. Prevalence of Plasmid-Mediated *ampC* Genes in Clinical Isolates of Enterobacteriaceae from Cairo, Egypt. *British Microbiology Research Journal*. **2013**, *3*, 525–537.

[10] Moland, E.S., Kim, S.Y., Hong, S.G., Thomson, K.S. Newer beta-lactamase: Clinical and Laboratory Implications, Part I. *Clinical Microbiology Newsletter*. **2008**, *30*, 71–77.

[11] Verdet, C., Benzerara, Y., Gautier, V., Adam, O; Ould-Hocine, Z., Arlet, G. Emergence of DHA-1-Producing *Klebsiella* spp. in the Parisian Region: Genetic Organization of the *ampC* and *ampR* Genes Originating from *Morganella morganii*. *Antimicrobial Agents and Chemotherapy*. **2006**, *50*, 607–617.

[12] Nevine, F., Doaa, G., Manal, E.S., Laila, A.F., Ehad, E.D., Soheir, E.A., Ashraf, S., Salwa, F., John, K. Detection of Plasmid-Mediated AmpC Beta-Lactamases in Clinically Significant Bacterial Isolates in a Research Institute Hospital in Egypt. *Life Science Journal*. **2013**, *10*, 2294–2304.

[13] Yoo, J.S., Byeon, J., Yang, J., Yoo, J.I., Chung, G.T., Lee, Y.S. High prevalence of extended-spectrum beta-lactamases and plasmid-mediated AmpC beta-lactamases in Enterobacteriaceae isolated from long-term care facilities in Korea. *Diagnostic Microbiology and Infectious Disease*. **2010**, *67*, 261–265.

[14] Mohamudha, P.R., Harish, B.N., Parija, S.C. Molecular description of plasmid-mediated AmpC β-lactamases among nosocomial isolates of *Escherichia coli* & *Klebsiella pneumoniae* from six different hospitals in India. Indian Journal of Medical Research. **2012**, *135*, 114–119.

[15] Tan, T.Y., Ng, S.Y., Teo, L., Koh, Y; Teok, C.H. Detection of plasmid-mediated AmpC in *Escherichia coli*, *Klebsiella pneumoniae* and *Proteus mirabilis*. Journal of Clinical Pathology. **2008**, *61*, 642–644.

[16] Woodford, N., Reddy, S., Fagan, E.J., Hill, R.L., Hopkins, K.L., Kaufmann, M.E., Kistler, J., Palepou, M.F., Pike, R; Ward, M.E; Cheesbrough, J., Livermore, D.M. Wide geographic spread of diverse acquired AmpC beta-lactamases among *Escherichia coli* and *Klebsiella* spp. in the UK and Ireland. *The Journal of Antimicrobial Chemotherapy*. **2007**, *59*, 102–105.

[17] Qin, X; Zerrt, D.M., Weissman, S.J., Englund, J.A., Denno, D.M., Klein, E.J., Tarr, P.I., Kwong, J., Stapp, J.R., Tulloch, L.G., Galanakis, E. Prevalence and mechanisms of broad-spectrum β-lactam resistance in *Enterobacteriaceae*: a children's hospital experience. Antimicrobial Agents Chemotherapy. **2008**, *52*, 3909–3914.

[18] Mai, M.H. and Reham, W. Phenotypic and Molecular Characterization of Plasmid Mediated AmpC β-Lactamases among *Escherichia coli*, *Klebsiella* spp., and *Proteus mirabilis* Isolated from Urinary Tract Infections in Egyptian Hospitals. *BioMed Research International*. **2014**, *2014*, 171548.

Correlation between natrium iodide symporter and c-fos expressions in breast cancer cell lines

A. Elliyanti
*Department of Medical Physics and Radiology, Faculty of Medicine, Universitas Andalas/
RS. Dr. M. Djamil, Padang, Indonesia*

N. Noormartany
Department of Clinical Pathology, Faculty of Medicine, Universitas Padjadjaran, Bandung, Indonesia

J.S. Masjhur
*Department of Nuclear Medicine, Faculty of Medicine, Universitas Padjadjaran/
RS. Dr. Hasan Sadikin, Bandung, Indonesia*

Y. Sribudiani, A.M. Maskoen & T.H. Achmad
Department of Biochemistry, Faculty of Medicine, Universitas Padjadjaran, Bandung, Indonesia

ABSTRACT: The aim of this study was to investigate the correlation between NIS and c-fos expression in breast cancer cell lines as a basis for radioiodine treatment. NIS and c-fos mRNA expressions were analyzed by reverse transcription PCR and their proteins were detected by immunocytofluorescence. We found an inverse correlation between NIS and c-fos expressions. NIS protein and mRNA were found only in SKBR-3 cells line. The protein was expressed in cytoplasm and membrane of cells. The copy number of mRNA NIS was 0.7 ± 0.01. C-fos protein and mRNA were found exclusively in MCF-7 cells. The copy number of mRNA c-fos was 1.13 ± 0.02. Our data suggest that SKBR-3 cell line seems to be suitable to receive radioiodine treatment. As c-fos expression can only be found in MCF-7, this finding has raised an assumption that c-fos may not be appropriate as a marker of proliferation in breast cancer.

1 INTRODUCTION

Natrium Iodide Symporter is a sodium/iodide co-transporter. It brings two sodium ions (Na^+) and the iodide (I^-) ion from the outside into the cell. (Oh, JR 2012 & Kogai, T 2012) Radioiodine has been used as adjuvant therapy for well-differentiated thyroid cancer. In addition to thyroid cancer, NIS expression is found mainly in invasive breast cancer cases that have poor prognosis. (Tazebay, UH 2011) Thus, they are likely to receive radioiodine as adjuvant therapy as seen in thyroid cancer.

The mechanism of NIS expression in breast cancer is still unclear. Several studies reported that, there were strong correlations between NIS expression and malignant transformation of human breast tissues. (Unterholzner, S 2006 & Tazebay, UH 2000) Cell proliferation is controlled by intracellular signaling pathways. Aggravation of the pathways is linked to malignant transformation. (Wagstaff, SC 2000) Furthermore, the proliferative signals are responsed by immediate early genes such as c-fos. It is a proto-oncogen and is regulated by intracellular signals events at multiple levels. Gee et al. study reported, there was correlation between increased c-fos expression and elevated cells proliferation and poor prognosis in breast cancer diseases. (Lu, C & Shen, Q 2005)

The objectives of this study were to investigate NIS and c-fos expressions in SKBR3 and MCF7 breast cancer cell lines and to determine the correlation between them. It can be used as a basic concept for radioiodine treatment in breast cancer.

2 MATERIALS AND METHODS

2.1 Cell lines culture

SKBR3 and MCF7 cell lines were used in this study. SKBR3 was obtained from the American Type Culture Collection (ATCC). MCF7 was kindly provided by Dr. Ahmad Faried from Faculty of Medicine, Universitas Padjadjaran, Bandung-Indonesia. MCF7 was cultured in RPMI 1640 medium. SKBR3 was cultured in Mccoy's 5A medium. Both cell culture mediums were supplemented with 10% fetal bovine serum, 1% Penicillin, 1% Streptomycin and 1% amphotericin B. The cells were incubated at 37°C and were supplied with 5% of carbon dioxide (CO_2) until 80% cell confluence.

2.2 RNA extraction

The cells were harvested at the appropriate time points by trypsinized, and then followed by centrifuge at 1000 rpm for a period of 4 minutes. The total RNA was isolated by using RNeasy mini kit (Qiagen #74106) according to the manufacturer's instructions and it was quantified using Nanodrop 2000.

2.3 Quantitative real-time reverse transcriptase—PCR (qRT-PCR)

Reverse transcription PCR was done on a Rotor Gene. Five ng of mRNA was reverse transcribed and analyzed by one step real-time quantitative PCR. (Quantitect probe RT-PCR Qiagen # 204443). NIS was amplified with forward primers: 5'-CCATCCTGGATGACAACTTGG-3', reverse: 5'-AAAAACAGACGATCCTCATTGGT-3', probe: 5'-AGAACTCCCCACTGGAAACAA-GAAGCCC (Presta 2005). C-fos was amplified with forward primers: GCGGACTAC-GAGGCGTCAT, reverse: GGAGGAGACCA-GAGTGGGC, probe: 5-/56-FAM/CTC CCC TGT/ZEN/CAA CAC ACA GGA CTT TTG C/31ABkFQ/-3 (Proudnikov 2003). Housekeeping gene (beta actin) was run parallel with the tested genes that were amplified with forward primers: 5-ACCGAGCGCGGCTACAG-3, reverse: 5-CTTAATGTCACGCACGAT TTC C-3, probe: 5-/56-FAM/TTCACCACC/ZEN/ACGGCCGAG c/31 ABkFQ/-3. Three independent qPCR assays were performed in triplicate. The expressions of the genes were reported as copy numbers.

2.4 Immunocytofluorescence

The cells were seeded on coverslips in 24-wells culture plates, then were fixated by 4% paraformaldehyde for 15 minutes. This was followed by incubation with protein blocking agent fluorescein-isothiocyanate (FITC) for 15 minutes. The cells were rinsed twice with an ice-cold PBS and followed by an overnight incubation with 2 µg/ml of rabbit polyclonal antibody anti-NIS (ab83816) and 5 µg/ml of rabbit polyclonal antibody anti-cfos (ab7963) at 4°C. Next, the cells were rinsed three times with PBS and incubated at room temperature with Goat anti-Rabbit IgG polyclonal secondary antibody, with dilution 1:1500 (ab 6716) for one hour. Again, the cells were rinsed three times with PBS. The coverslips were placed over and mounted with fluoroshield containing DAPI. The slides were inspected under immunofluoresense microscope (Olympus BX51) with 200x magnification. The cells which were only incubated with secondary antibody were used as negative control.

2.5 Statistical analysis

Data were shown as mean of three individual experiments and presented as mean ± SD. Pearson correlation was used to test the correlation between NIS with c-fos expressions.

3 RESULTS

3.1 RNA extraction

The expressions of NIS mRNA were only detected in SKBR-3. Meanwhile, c-fos mRNA was detected solely in MCF-7 cells as show in Figure 1.

3.2 mRNA expressions of NIS and c-fos

We analyzed the expressions of NIS *mRNA* in SKBR3 and MCF-7 cell lines. We observed that NIS expression only shown in SKBR-3 cells, with the copy number of NIS mRNA was 0.7 ± 0.01. We cannot identify its expressions in MCF-7 cells. On the other hand, we detected c-fos mRNA expression in MCF-7 cells but not in SKBR-3 cells. The copy number of mRNA c-fos was 1.13 ± 0.02.

3.3 NIS and c-fos protein

Results of immunofluorescence showed that NIS proteins were expressed in SKBR3 cell line as shown in Figure 2. The NIS expressions were predominantly seen at cytoplasm instead of at cell membrane. On the other hand, we cannot detect NIS protein expressions in MCF7 cell lines. This finding has an agreement with qRT-PCR results.

In contrast to NIS protein, c-fos is expressed only in MCF7 cells and seen at nuclei as shown in Figure 3.

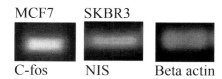

Figure 1. Expressions of c-fos in MCF-7 and NIS in SKBR-3 cells were investigated by qRT-PCR. A predicted 231-bp c-fos, 100-bp NIS and 80 bp beta actin PCR products were quantified by Nanodrop.

Figure 2. Immunocytofluorescence stains show NIS protein (green) mainly in cytoplasm (arrow) and membrane (head arrow). Nuclei is (blue) at 200x magnification from SKBR3 cell line. (a) merge FITC+DAPI. (b) DAPI. (c) FITC.

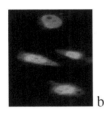 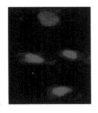

Figure 3. C-fos protein expressions at nuclei showed as green staining by FITC and nuclei as blue staining by DAPI at 200x magnification in MCF-7 cells.

3.4 Correlation between NIS with c-fos expressions

SKBR3 cell lines has NIS expression. It represents HER2 subtype. In contrast with NIS expression, c-fos is expressed in MCF7 which represents luminal A subtype. From this study, we found an inverse correlation between NIS and c-fos expressions.

4 DISCUSSION

The expression of NIS in extra-thyroid tissues has been reported in some previous studies. These findings have been considered for radioiodine therapy in extra-thyroid tumor, including breast cancer. A study by Tazebay et al, reported that NIS was expressed in more than 80% of invasive breast cancer tissues. (Tazebay, UH 2011) Another study by Wapnir et al, reported that the expression was found in 76% of invasive breast cancer tissues. (Wapnir, IL 2003) Studies by other groups reported that 34% out of 44 breast cancer tissues expressed NIS and was also reported that 65.5% out of 23 triple negative breast cancer tissues expressed NIS. (Reiner, C 2009)

In this study, we found that NIS *mRNA* and protein expressions were identified only in SKBR3 cell. SKBR3 is a breast cancer cell line with HER2 receptor positive (HER2+), estrogen and progesterone receptors negative (ER−/PR−). (Holliday, DL 2011 & O'Brien, KM 2010) Furthermore, the expressions of NIS were not detected in MCF7 cells. This cell line is luminal A type of breast cancer model, with estrogen and progesterone receptors positive (ER+/PR+) and HER2 receptor negative (HER2−). (Holliday, DL 2011 & O'Brien, KM 2010) The previous studies reported that, Retinoid Acid (RA) treatment is mandatory to induce NIS expression in MCF7 cell. (Dohan, O 2006 & Kogai, T 2006) Thus, it is assumed that NIS expression in MCF7 is RA dependent. Furthermore, the efficacy of radioiodine therapy for breast cancer will be likely depended on the level of NIS expression.

It is important to determine molecular breast cancer subtypes to be a candidate for radioiodine treatment. Radioiodine seems to be a promising adjuvant treatment for breast cancer, mainly for cases which are unresponsive to the treatment. Several agents are used to increase NIS expression, such as retinoic acid, hormones and proliferative agents. (Kogai, T 2012 & Jung, KH 2008 & Arturi, F 2005)

A study by Tazebay et al reported, that there was a correlation between NIS expression and malignant transformation of human breast tissue. (Unterholzner, S 2006 & Tazebay, UH 2000) Furthermore, our study showed SKBR3 cell lines which represent HER2 subtype were expressing NIS endogen. This leads us to assume that increasing of cell proliferation will up-regulate NIS expressions in breast cancer.

C-fos is a proto-oncogene. It plays a critical regulator in response to the proliferative signals. (Mahner, S. et al. 2008 & Middle-langosch, K 2004 & Wagstaff, SC 2000) It is used as a marker of proliferation. (Jimeno, A 2006) Furthermore, c-fos also has control in breast cancer cells growth regulation. (Lu, C & Shen, Q 2005) In our study, we found c-fos mRNA and protein expressions only in MCF 7 cells. Conversely, the expressions were absent in SKBR3 cells. The growth of Estrogen Receptor (ER) positive breast cancer cells are effectively inhibited by Tam67 but not the ER negative breast cancer cells. It can be hypothesized that an absence of c-fos expression in SKBR3 cells are related to ER negative cell lines type. (Lu, C & Shen, Q 2005)

5 CONCLUSIONS

The expression of NIS varies among breast cancer cell lines. In contrast to SKBR3 which represents a HER2 subtype, MCF7 cell lines that represent a luminal A subtype do not expressed NIS. The correlation of expression between NIS and c-fos is inverted. This finding has raised an assumption that c-fos may not be an appropriate marker of proliferation in breast cancer. As SKBR3 cell lines has NIS expression, therefore, radioiodine could be potential for breast cancer adjuvant therapy, particularly for HER2.

ACKNOWLEDGEMENTS

This work was accomplished by a grant from Indonesia Risbin Iptekdok Project 2013.
Thank you very much to Dr. Andani Eka Putra who help trouble shooting of qPCR and Dewi Rusnita MD, MSc who assist a manuscript preparation.

CONFLICT OF INTEREST STATEMENT

No potential conflicts of interest are disclosed.

REFERENCES

Arturi, F. et al. 2005. Regulation of Iodide Uptake and Sodium/Iodide Symporter Expression in the MCF-7 Human Breast Cancer Cell Line. J Clin Endocrinol Metab 90(4):2321–2326.
Dohan, O. et al. 2006. Hidrocortisone and Purinergic Signaling Stimulate Sodium/Iodide Symporter (NIS)-Mediated Iodide Transport in Breast Cancer Cells. Mol Endocrinol 20(5):1121–1137.
Holliday, DL. & Speirs, V. 2011. Choosing the Right Cell Line for Breast Cancer Research. Breast Cancer Research. BioMed Central 215.
Jimeno, A. et al. 2006. C-fos Assesment as a Marker of Anti-Epidermal Growth Factor Receptor Effect. Cancer Res 66(4):2385–2390.
Jung, KH. et al. 2008. Mitogen-Activated Protein Kinase Signaling Enhances Sodium Iodide Symporter Function and Efficacy of Radioiodine Therapy in Nonthyroidal Cancer Cells. J. Nuclear Medicine 49:1966–1972.
Kogai, T. et al. 2006. Enhancement of sodium/iodide symporter expression in thyroid and breast cancer. Endocrine-Related Cancer 13:797–826.
Kogai, T. et al. 2012. Regulation of Sodium Iodide Symporter Gene Expression by Rac1/p38β Mitogen-activated Protein Kinase Signaling Pathway in MCF7 Breast Cancer Cells. J. Bio. Chem 287:3292–3300.
Lu, C. & Shen, Q. 2005. C-fos is critical for MCf-7 Breast Cancer Cell Growth. Oncogene 24:6516–6524.
Mahner, S. et al. 2008. C-fos Expression is a Molecular Predictor of Progression and Survival in epitelial Ovarian Carcinoma. British Journal of Cancer 99:1269–1275.
Milde-Langosh, K. et al. 2004. The Role of the AP-1 Transcription Factors c-fos, fosB, Fra-1 and Fra-2 in the Invasion process of Mammary Carcinomas. Breast Cancer Research and Treatment 86:139–152.
O'Brien, KM. et al. 2010. Intrinsic Breast Tumor Subtypes, Race, and Long-Term Survival in the Carolina Breast Cancer Study. Clin Cancer Res 16(24):6100–6110.
Oh, JR. & Ahn, BC. 2012. False Positive Uptake on Radioiodine Whole Body Scintigraphy Physiologic and Pathologic Variants Unrelated to Thyroid Cancer. Am J Nucl Med Mol Imaging 2(3):362–385.
Presta, I. et al. 2005. Recovery of NIS Expression in Thyroid Cancer Cells by Overexpression of Pax8 Gene. BMC cancer 2005(5):80.
Proudnikov, D. et al. 2003. Optimizing primer-probe design for fluorescent PCR. J. of Neuroscience Methods 123(1):31–45.
Reiner, C. et al. 2009 Endogenous NIS Expression in Triple-Negative Breast Cancers. Annals of Surgery Oncology 16(4):962–968.
Tazebay UH. 2011 Regulation of the Functional Na$^+$/I$^-$ Sumporter (NIS) Expression in Breast Cancer Cells. In Breast Cancer-Recent advances in Biology, Imaging and Therapeutics. In Tech 103.
Tazebay, UH. & Wapnir, IL. 2000. The Mammary Gland Iodine Transporter is Expressed During Lactation and in Breast Cancer. Nature Medicine 6:871–878.
Unterholzner, S. & Willhauck, M J. 2006. Dexamethasone Stimulation of Retinoic Acid-Induced Sodium Iodide Symporter Expression and Cytotoxicity of 131-I in Breast Cancer Cells. The J of Clin Endocrinol Metab 91(1):69–78.
Wagstaff, SC. et al. 2000. Extracellular ATP Activites Multiple Signalling Pathways and Potentiates Growth Factor-Induced c-fos Gene Expression in MCF-7 Breast Cancer Cells. Carcinogenesis 21:2175–2181.
Wapnir, IL. et al. 2003. Immunohistochemical profile of the sodium/iodide symporter in thyroid, breast and other carcinoma using high density tissue microarrays and conventional section. J. Clincal Endocrinology and Metabolism 88:1880–1888.

Fat mass profile in early adolescence: Influence of nutritional parameters and rs9939609 FTO polymorphism

S.N. Fatimah, A. Purba, K. Roesmil, G.I. Nugraha & A.M. Maskoen
Faculty of Medicine, Padjadjaran University, Bandung, Indonesia

ABSTRACT: Fat mass is one of the important parameters for metabolic risk screening in adolescents. A genetic variant of FTO, such as rs9939609, is predisposed to higher fat mass. Nutritional intake in adolescence is an important factor to optimize growth spurt and maintain health status. The objective of this study was to identify the characteristics of fat mass associated with rs9939609 FTO polymorphism, macronutrient intake, and nutritional status parameters in early adolescence. A total of 192 early adolescents aged 10–14 years were genotyped for rs9939609 polymorphism, underwent body composition and anthropometric measurements, and were asked about their eating pattern of macronutrients. A bivariate analysis showed significant differences in fat mass based on gender (p: 0,003), height (p: 0,05), and body mass index (p: 0,00), and non-significant differences based on rs9939609 FTO polymorphism (p: 0,89), energy intake (P: 0,94), and protein intake (p: 0,2). The result indicated that gender, BMI, height, and protein intake had an influence on fat mass. Hormonal influence in early adolescence increased fat deposition in females. Body height affected body composition because of differences in skeletal and muscle mass. There was no influence of the rs9939609 FTO gene variant between races, although FTO had the highest risk allele among other genetic factors predisposed to fat deposition. Genetic and environmental factors had an influence on fat mass. As a result, it became important to analyze other obesity-related genetic and environmental factors because many indirect correlations were found between chronic disease, genetic characteristic, adolescent sedentary lifestyle, and long-term comorbidity.

Keywords: fat mass, early adolescence, height, BMI, nutritional intake, 9939609 FTO polymorphism

1 INTRODUCTION

Body fat mass describes adequate nutrient reserve, optimal growth, metabolic conditions, disease susceptibility, and various risk factors.[1-4]

Fat mass is influenced by genetic and environmental factors. Thus, it is important to assess whether genetic factors can increase risk factors associated with adolescent body composition.[5,6] Fat mass obesity (FTO) gene is a gene that encodes the FTO protein.[6,7] This protein functions as a co-substrate of 2-oxoglutarate which is an intermediate protein in the citric acid cycle.[7,8] The FTO gene interacts with other genes that affect food intake by regulating hunger and satiety. Various studies have suggested that FTO has a significant role in fat mass increase and BMI.[9]

FTO gene polymorphism tends to increase body fat mass. A study on SNP rs9939609 polymorphism of the FTO gene conducted in Pakistani women showed a significant association with increased BMI and the risk of obesity.[10,11]

The adequacy of energy and protein intake is important to maintain the continuity of tissue hyperplasia and hypertrophy. Lack of nutrition during growth, especially during the catch-up growth, correlates with the occurrence of short stature. Low intake of nutrients can occur primarily due to the lack of nutrition or secondarily because of increasing need, for example, in children with chronic infection.[12,13]

The problem of malnutrition occurs in all age groups, including teenagers and adolescents. For example, the data from Riset Kesehatan Dasar (RISKESDAS) showed that nutrient intake profiles in West Java are still low. Mean energy and protein intake in West Java is still below the recommended levels.[14]

The objective of this study was to identify the characteristics of fat mass associated with rs9939609 FTO polymorphism, macronutrient intake, and nutritional status parameters in early adolescence.

2 MATERIALS AND METHODS

2.1 Sample

The present cross-sectional study was conducted from September 2014 until April 2015 in Sub-district Jatinangor, which is a semi-urban area in

West Java province with high dynamic changes in lifestyle. A total of 192 subjects, aged between 10 and 14 years, were selected from three elementary schools and four junior high schools.

2.2 Anthropometric measurements

All the children underwent a series of anthropometric measurements, which included height, weight, BMI, and body composition. The measurements were taken according to the standard protocols. Standing height, without shoes, was measured (to the nearest 0.1 cm) with a stadiometer (Leicester Stadiometer, Seca). Body weight, in light clothing, was measured (to the nearest 0.1 kg) with a mechanical floor scale (Tanita). The BMI was calculated as weight (in kilograms) divided by the square of height (in meters).

2.3 Body fat mass

Fat mass was measured by using a bioimpedance analyzer (Tanita SC 240 MA). The procedure for measuring weight and body fat is as follows: turn on the scale, set the date of birthday, select female or male, press specify height and then press the set button. Next, step on to the scale and measure the weight and body fat after "0.0" is shown on display within 30 seconds. The readings will be shown for 40 seconds. The unit will then shut off and the readings will be stored in the memory.

2.4 Measurement of food intake

Food intake was collected using the 2×24 hour recall. Subjects were asked to report all of the foods eaten or drunk on the previous day in an uninterrupted free flowing list. For each item of food or drink in the quick list, the respondents were asked to provide additional details, including: the time at which the food or drink was consumed, a full description of the food or drink, including brand name where available, any food likely to be eaten in combination, e.g. milk in coffee, the quantity consumed, based on household measures and food model of different portion sizes of food, or actual weights from labels or packets. The interviewer reviewed all of the food eaten and drunk in chronological order, prompting for any additional eating or drinking occasions and food or drinks consumed.

Nutrient composition was analyzed according to the Indonesian Food Composition Table (SPSS, and Excel Program from Microsoft).

2.5 DNA extraction and genotyping

Genomic DNA was extracted from whole blood using a genomic DNA purification kit (Fermentas, USA). Genotyping of rs9939609 polymorphism at the *FTO* locus was carried out by the Polymerase Chain Reaction–Restriction Fragment Length Polymorphism assay (PCR-RFLP). A DNA fragment containing rs9939609 polymorphism was amplified using specific primers (forward primer sequence: AACTGGCTCTT-GAATGAAATAGGATTCAGA and reverse primer sequence: AGAGTAACAGAGACTATC-CAAGTGCAGTAC). The PCR was carried out using a thermocycler (Icycler 5; BioRad, USA) under optimized conditions. In a 25 μL reaction, PCR components consisted of 50 ng DNA, 1X Taq buffer, 2 mM MgCl$_2$, 200 μM of each dNTP, 10 ρmol of each primer, and 1U *Taq* DNA polymerase. Thermal cycling was performed as follows: initial denaturation at 95°C for 4 min, followed by 35 cycles of denaturation at 94°C for 30 sec, annealing at 58°C for 30 sec and extension at 72°C for 1 min, and then a final extension step at 72°C for 10 min. Amplified products were digested with the restriction enzyme *ScaI* (Favorgen, Taiwan) to analyze polymorphism by RFLP assay.

2.6 Statistical analysis

Data were analyzed using SPSS software (version 20, SPSS). Quantitative traits (height, weight, BMI, fat mass, and nutrient intake) were analyzed by univariate analysis of variance. Age and sex were used as covariates when appropriate. Descriptive statistics were used to determine the mean and Standard Deviation (SD) of age- and gender-specific anthropometric indicators. Fisher's exact test was used to determine the significant difference between fat mass and another variable. The level of significance was set at $P < 0.05$. Table 1 presents the means and SD and t-ratios of absolute body mass and body composition of both the normal and malnourished groups of children aged 3 to 11 years.

3 RESULTS

A total of 192 early adolescents participated in the study. The characteristics of the subjects including

Table 1. Anthropometric characteristics of the subjects.

Variable	Distribution (n: 192)
Gender	
Male	41,14%
Female	58,86%
Age (mean, SB)	12,02 (1,22)
Height (mean, SB)	143,40 (9,40)
Weight (mean, SB)	37,05 (9,42)
BMI (mean, SB)	18,54 (3,24)

age, gender, height, weight, and BMI are listed in Table 1. Their anthropometric characteristics are presented in Table 1.

Among the gender characteristics, male early adolescents were more frequent than female adolescents. The height and weight data showed a wide distribution range.

The characteristics of fat mass given in Table 2 describe the wide distribution range from 0,8 to 31 kg.

Furthermore, Table 3 reports the characteristics of the FTO rs9939609 polymorphism. It shows that the frequency of homozygote and heterozygote mutants is less than the homozygous wild type (17, 17%).

The characteristics of the subjects based on total energy intake (kcal), protein (g), fat (g), and carbohydrates (g) presented in Table 4 show large variation, although the repetition of the 24 hour recall showed a homogeneous meal pattern.

Association between fat mass and each variable.

A bivariate analysis showed significant differences in fat mass based on gender, height, and body mass index, and non-significant differences based on rs9939609 FTO polymorphism, energy intake, and protein intake.

Overall, the result from the multivariate analysis indicated that gender, BMI, height, and protein intake could predict fat mass. Age, gender, body height, BMI, and protein intake could predict fat mass with moderate power.

Table 2. Fat mass characteristics of the subjects.

Distribution	(n:192)
Mean (SD)	7,17 (5,45)
Median (min–max)	6,3 (0,8–31)

Table 3. FTO rs9939609 polymorphism genotype characteristics of the subjects.

Genotype	Frequency (n:192)
AA/AT	33 (17,18)
TT	159 (82,81)

Table 4. Nutrient intake characteristics of the subjects.

Variable (n:192)	Mean (SD)	Median (min–max)
Energy (kcal)	1616,83 (598,68)	1518,86 (461,94–3597,71)
Protein (g)	44,71 (18,37)	43,47 (10,88–101,19)
Fat (g)	60,03 (81,83)	49,85 (2,75–111,70)
Carbohydrate (g)	872,2 (117,44)	235,29 (393–1627,60)

Table 5. Association between fat mass and variables.

Variable	Mean (SD)	P
Gender**		0,003
Male	41,14%	
Female	58,86%	
Height*	143,40 (9,40)	0,04
BMI*	18,54 (3,24)	0,0001
Energy*	1616,83 (598,68)	0,94
Protein*	44,71 (18,37)	0,2
Fat*	60,03 (81,83)	0,36
Carbohydrate*	872,2 (117,44)	0,43
Polymorphism**		0,89
AA/AT	33 (17,18)	
TT	159 (82,81)	

*Chi square; **Fisher's exact.

Table 6. Association between fat mass and variables.

Variable	p	R2
Height	0,47	
Gender	0,17	
Age	0,02	0,4
BMI	0,00	
Protein intake	0,03	

4 DISCUSSION

In early adolescence, hormonal influence increased fat deposition in females. The reason is that this condition may occur due to differences in the influence of reproductive hormones in the age range of 10 to 14 years, and the growth rate difference between boys and girls.[12]

Body height affects body composition because of differences in skeletal and muscle mass.[13]

BMI was in the borderline normal range, and a clinical significant distribution of nutritional status is lacking.

The result of fat mass distribution shows considerable variation. This would likely influence reproductive hormones that differ in male and female early adolescents, because in male adolescents, the focus is on the growth of fat-free mass, while in women, higher fat mass deposition supports the maturation of the reproductive function.[14]

Furthermore, the results indicated that this gene is associated with the control of energy metabolism, physical activity, and eating behavior. Research on mice showed a statistically significant association between the FTO gene and increased expression of the protein intake. FTO rs9939609 gene polymorphism showed that the frequency of polymorphism is not as high as other races such as the research on the profile of the rs9939609 FTO gene polymorphism in Europe and Japan. There is no influence of the rs9939609 FTO gene variant with the highest risk allele among other genetic factors on fat deposition.[15]

Although FTO alleles have the highest frequency among other genes that affect fat deposition, it does not support adequate nutrition as a substrate for fat deposition.[15]

The distribution of the nutrition profile reported that nutritional intake is lower than the Indonesian reference nutrient intake. These characteristics are similar to some research conducted in developing countries that show a correlation between nutrition and retarded linear growth.[16]

Multivariate analysis with linear regression shows that several factors affect fat deposition. In this study, age is inversely correlated with fat mass. However, it is necessary to study other factors that affect fat deposition due to the effect of growth hormones on fat mass and fat-free mass, especially against smaller mass with older age.

Body height correlates with increased fat mass, which is different from the results of previous studies where the height is inversely associated with fat mass; however, other factors that affect fat deposition need to be considered.

There is a relationship between fat mass and BMI, which is in line with other studies, and it can be predicted that BMI and fat mass have the same predictive clinical interpretation of the risk factors for diseases associated with fat mass.

Adequate protein intake decreases fat deposition. Protein intake analyzed in this study can be considered as a possible risk factor for excess fat deposition. This study has several strengths. First, this survey is comprehensive and measures a broad range of nutrition-related parameters.

Second, it provides a valuable database and adds to the scanty literature on the regional nutritional status.

A limitation of this study is that due to the different sampling protocol, our study was sampled at the school and not at the household level.

Another limitation is the overestimation of portion sizes in dietary assessment, as the 24-hour recall was proxy-reported by the parents of the subjects. Some subjects had difficulty in estimating the portion sizes of the foods consumed.

5 CONCLUSION

This study provided a significant amount of data on the nutritional status and variables related to early adolescence.

Collectively, the findings of the present study reveal that there is an association between environmental factors and fat mass but no association with rs9939609 FTO gene polymorphism.

A strategy for improving the normal deposition of fat mass is to increase nutritional intake, especially protein intake.

Many multifactorial genetic and environmental factors have influenced fat mass. Therefore, it is important to analyze, for example, physical activity and morbidity status, and another genetic factor because many indirect correlations have been found between chronic disease, genetic characteristic, trend of adolescent sedentary lifestyle, and long-term comorbidity.

ACKNOWLEDGMENTS

The study was funded by the Progress Funding and Finishing Doctoral Study Funding for Lecturer from Padjadjaran University. The authors thank all the subjects, teachers, supervisors, and parents for their participation and cooperation during data collection. The research team, and all those who are involved in this project are acknowledged and much appreciated for their effort and dedication. The authors are also grateful to Prof. Ambrosius Purba, Dr Gaga Irawan Nugraha, Prof. Kusnandi Roesmil, and Dr Ani Melani Maskoen for their support and assistance in data analysis.

REFERENCES

[1] Gibson. R. *Principles of Nutritional Assessment.* NewYork. Oxford University Press. 2005.
[2] Eboh LO., Boye TE. Investigation of body composition of normal and malnourished rural children (3–11 Years) in the Niger—Delta Region of Nigeria. Pakistan J of Nut 4 2005; (6):418–422.
[3] Cesani M.F, Garraza M., Sanchís ML., Luis MA., Torres MF., Quintero FA, et al. A Comparative Study on Nutritional Status and Body Composition of Urban and Rural School children from Brandsen District (Argentina) *PLOS ONE 2013*; 8: e52792.
[4] Jafar TH., Qadri Z., Islam M., Jatcher J., Bhutta ZA., Chaturvedi Z. Rise in childhood obesity with persistently high rates of undernutrition among urban school-aged Indo-Asian children. Arch Dis Child 2008; 93:373–378.
[5] Loos RJ. Recent progress in the genetics of common obesity 2009 Br J Clin Pharmacol 68:6:811–829.
[6] Day F., Loos RJ. Developments in Obesity Genetics in the Era of Genome-Wide Association Studies J Nutrigenet Nutrigenomics 2011; 4:222–238.
[7] Williams KW., Elmquist JK. From neuroanatomy to behavior: central integration of peripheral signals regulating feeding behavior. Nature Neuroscience. 2012; 15:10.

[8] Sylvain Sebert S., Tuire Salonurmi T., Keinänen-Kiukaanniemi S., Savolainen M., Herzig KH., Symonds ME. Programming effects of FTO in the development of obesity (in press).

[9] Fischer et al. FTO effect on energy demand versus food intake Arising from. Nature 2009; 458:894–898.

[10] Cecil, JE., Tavendale, R., Watt, P., Hetherington, MM., Palmer, CNA. An Obesity-Associated *FTO* Gene Variant and Increased Energy Intake in Children. N Engl J Med. 2008; 359:2558–66.

[11] Liu G, Zhu H, Lagou V, Gutin B, Stallmann-Jorgensen I, Treiber F, et al. *FTO variant rs9939609* is associated with body mass index and waist circumference, but not with energy intake or physical activity in European- and African-American youth R. *BMC Medical Genetics* 2010; 11:57.

[12] Cromer B. Adolescent development In: Kliegman RM, Behrman RE, et al. editors. Nelson's Textbook of Pediatrics. 18 ed. Philadelphia: Saunders Elsevier; 2011; p. 649–659.

[13] Hills AP., King NA., Armstrong TP. The Contribution of Physical Activity and Sedentary Behaviours to the Growth and Development of Children and Adolescents Implications for Overweight and Obesity Sports Med. 2007; 37(6):533–545.

[14] Riset Kesehatan Dasar. Badan Penelitian dan Pengembangan Kesehatan Kementerian Kesehatan RI. 2013.

[15] Sterling R., Miranda J., Gilman RH., Cabrera L., Sterling CR., Bern C., et al. Early Anthropometric Indices Predict Short Stature and Overweight Status in a Cohort of Peruvians in Early Adolescence. 2012. Am.J.Physical Anthropol 148:451–461.

[16] Motswagole BS, Kruger HS, Faber M, Senior Monyeki KD. Body composition in stunted, compared to non-stunted, black South African children, from two rural communities S Afr J Clin Nutr. 2012; 25(2):62–66.

[17] Gulati P., Yeo GSH. The biology of FTO: From nucleic acid demethylase to amino acid sensor. Diabetologia 2013; 56:2113–2121.

[18] Ayoola O, Ebersole K, Omotade O, Tayo BO, Brieger WR, Salami K, et al. Relative Height and Weight among Children and Adolescents of Rural Southwestern Nigeria Ann Hum Biol. 2009; 36(4): 388–399.

A study of the palatal rugae pattern as a bioindicator for forensic identification among Sundanese and Malaysian Tamils

R. Khaerunnisa
Biomedical Sciences Program, Graduate School of Medical Sciences, Universitas Padjadjaran, Bandung, West Java, Indonesia

M. Darjan & I.S. Hardjadinata
Department of Oral Biology, Faculty of Dentistry, Universitas Padjadjaran, Bandung, West Java, Indonesia

ABSTRACT: The use of human palatal rugae has been suggested as an alternative method for forensic identification, especially if teeth are lost due to any reason. This study was conducted to identify and compare the palatal rugae pattern of Sundanese and Malaysian Tamil students at the Faculty of Dentistry, Universitas Padjadjaran Indonesia. A total of 60 subjects comprising 30 Sundanese and 30 Malaysian Tamils aged between 18 and 25 years were included in this study. Rugae patterns were analyzed based on the Thomas and Kotze classification. The chi-square test was used to study the statistical significance. Sundanese had more wavy pattern (70%), which was significantly higher in proportion ($P < 0.05$) when compared with Malaysian Tamils (20%), while Malaysian Tamils had predominantly a curved pattern (66.7%) and showed significant differences in proportion ($P < 0.05$) when compared with Sundanese (16.6%). The difference in the palatal rugae pattern between the two different ethnic groups can be used as a bioindicator for population identification.

Keywords: Palatal rugae pattern, population identification, Sundanese, Malaysian Tamils

1 INTRODUCTION

In forensic identification, establishing a person's identity can be a very difficult process. Fingerprints, DNA profiling, and odontology are the most common techniques used to make a positive human identification. As it cannot be always used in several situations, other techniques can be successfully used as an alternative in human identification. The use of human palatal rugae in forensic dentistry has been suggested to establish a person's identity. It can be considered as an alternative method in human identification if teeth are lost due to any reason, the most common of which is trauma and also in edentulous cases (Caldas et al. 2007, Bansode & Kulkarni 2009).

Palatal rugae are the anatomical folds that are located in the anterior third of the palate behind the incisive papilla. It is formed in the early intrauterine life from the hard connective tissue covering the palatal bone (Bansode & Kulkarni 2009). The patterns are formed by the third month of prenatal life and remain stable throughout the person's life. Once it is formed, it only changes in length due to normal growth.

Palatal rugae are guarded by other oral structures such as the teeth, bone, buccal pad, cheek, tongue, and lips. Owing to its internal position in the oral cavity, palatal rugae are protected from trauma, high temperatures, and decomposition after death up to 7 days (Segelinck et al. 2005, Mutsuhubramanian et al. 2005, Acharya et al. 2007).

Palatal rugae pattern may also be specific to racial groups, allowing population identification. Studies have shown that certain palatal rugae patterns are particular to different populations. Kapali et al. (1997) studied rugae patterns in different ethnic groups of Australian Caucasians and Aborigines, and revealed a statistically significant association between rugae shapes and ethnicity. Shetty et al. (2005) carried out a comparative analysis of palatal rugae pattern in Mysorean and Tibetan populations, and revealed a difference in the length and shape of rugae between the two tribes. Thus, the uniqueness to different ethnic groups, stability, postmortem resistance, and additionally low utilization cost makes palatal rugae useful for population identification in forensic dentistry (Nayak et al. 2007).

This study was conducted to identify and compare the palatal rugae pattern of Sundanese and Malaysian Tamil students at the Faculty of Dentistry, Universitas Padjadjaran Indonesia. These groups were selected for the study due to racial differences. Sundanese are ethnic groups native to the western part of the Indonesian Island of Java. Racially, Sundanese belong to the Southern Mongoloid. While Malaysian Tamils are Tamil people born in or immigrated to Malaysia. Tamils are Caucasoid, a part of Dravidian branch, which is largely a part of the Mediterranean branch of the Caucasoid race.

2 MATERIALS AND METHODS

In this study, 60 subjects were randomly selected from Faculty of Dentistry Universitas Padjadjaran Indonesia students, comprising 30 Sundanese and 30 Malaysian Tamils aged between 18 and 25 years. Before conducting the study, the subjects were informed about this study and informed consent was obtained from each participant. The study protocol was reviewed and approved by the Ethical Committee of the Institution.

Subjects with any palatal abnormalities and deformity such as cleft palate, bony protuberances, scars and palatal asymmetries, previous history of any trauma and surgery involving palate, allergic to impression material, having active lesions in the palate, wearing partial dentures, and orthodontic appliance were excluded.

From each subject, an alginate impression of the maxillary arch was made using an appropriate tray, and the cast was prepared using Type III dental stone. The palatal rugae patterns were highlighted using a black graphite pencil on the dental cast (see Figure 1) and were analyzed according to Thomas

Figure 1. Palatal rugae pattern analysis using dental cast of subjects.

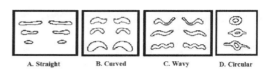

Figure 2. Palatal rugae classification by Thomas and Kotze.

and Kotze classification (Paliwal et al. 2010) (see Figure 2). This classification divided the palatal rugae pattern into four types based on its shape: straight, curved, wavy, and circular. Straight type ran directly from their origin to termination in a straight line. The curved type had a simple crescent shape which curved gently. Wavy rugae had a slight curved at the origin or termination and rugae that formed a definite continuous ring formation were classified as circular.

All the identifications were carried out by a single examiner. The association between palatal rugae pattern and ethnicity was tested using the chi-square analysis.

3 RESULTS

The major palatal rugae patterns in the total subjects of Sundanese and Malaysian Tamils are presented in Table 1. The more common pattern was wavy (70%), followed by curved (16.6%), straight (6.7%), and circular (6.7%) in the Sundanese population, whereas in the Malaysian Tamil population, curved (66.7%) pattern was more common followed by wavy (20%), straight (10%), and circular (3.3%).

The most common patterns in both ethnic groups were wavy and curved forms, whereas straight and circular types were least common. Overall, there was a statistically significant association between the palatal rugae pattern and the ethnicity (chi-squared value = 17.87 with three degrees of freedom, $P < 0.05$) (see Table 1). In Sundanese population, the most common pattern was the wavy one followed by the curved one. On the other hand, the most predominant pattern in Malaysian Tamil students was the curved one followed by the wavy one.

The Sundanese had a wavier pattern with percentage values of 70%, which is significantly higher in proportion ($P < 0.05$) when compared with the Malaysian Tamils with percentage values of 20%, while Malaysian Tamils proved to be curved predominant with percentage values of 66.7% and showed significant difference in proportion ($P < 0.05$) when compared with the Sundanese with percentage curve values of 16.6%.

Table 1. Distribution of different palatal rugae patterns in total subjects of Sundanese and Malaysian Tamils*.

Shapes	Sundanese n (%)	Malaysian Tamils n (%)
Curved	5 (16.6)	20 (66.7)
Wavy	21 (70)	6 (20)
Straight	2 (6.7)	3 (10)
Circular	2 (6.7)	1 (3.3)

* Chi-squared value = 17.87 with three degrees of freedom, $P < 0.05$.

4 DISCUSSION

Palatal rugae first appeared in an anatomy textbook by Winslow in 1732, and was first illustrated by Santorini in 1975 (Shetty et al. 2005). The use of palatal rugae was first suggested as a method of human identification by Harrison Allen in 1889 (Caldas et al. 2007). Since then, many have attempted to study the palatal rugae in order to establish the identity of an individual.

The palatal rugae are irregular and asymmetrical anatomical fold that are located in the anterior third of the palate behind the incisive papilla and across the anterior part of the median palatal raphe. It is formed in the early intrauterine life from the hard connective tissue covering the palatal bone and its formation is under genetic control (Bansode & Kulkarni 2009).

Many researchers hypothesize that, once it is formed, the palatal rugae pattern is distinct to an individual. Its design and structure are unchanged and are not altered by chemicals, physicals, thermal factors, or diseases. It does not change throughout the person's life, except for an increase in length due to normal growth.

The potential of palatal rugae in forensic identification has advantages because of its simplicity, cost-effectiveness, and reliability (Nayak et al. 2007). There are several techniques for analyzing the palatal rugae. The study of dental cast for rugae pattern is the most simple and reliable method.

This study was conducted to identify and compare the palatal rugae pattern in two different populations. Rugae patterns were analyzed according to Thomas and Kotze classification. This method was simple to perform and less time-consuming. The total numbers of palatal rugae in the Sundanese and Malaysian Tamil groups were found to be different from one another and they show statistically significant difference. These findings were consistent with those of previous studies conducted by Hauser et al. (1989) in Swazi and Greek populations, by Kapali et al. (1997) in Australian Caucasians and Aborigines also by Shetty et al. (2005), and Paliwal et al. (2010) among Indian population. These various authors found definite differences in the palatal rugae patterns between two populations.

Analysis of the palatal rugae pattern showed the wavy pattern to be more prevalent in Sundanese population, followed by the curved one, whereas the curved pattern was found to more prevailing in Malaysian Tamil population followed by the wavy one. It was also noticed that straight and circular types were least common in both populations. Our results are concordant with the data revealed by various authors in the similar studies conducted earlier (Kapali et al. 1997, Shetty et al. 2005). Preethi et al. (2007) also reported similar findings on Western and South Indian populations, where circular types were few in number (Nayak et al. 2007).

The significant differences in palatal rugae patterns among Sundanese and Malaysian Tamils have stated the fact that certain patterns are more common in certain populations, which emphasized its importance in population identification.

5 CONCLUSIONS

This study concludes that the wavy pattern was more common in Sundanese, whereas the curved pattern was more common in Malaysian Tamils. The fact that certain patterns are more common in certain populations emphasized that it can be used as a bioindicator for population identification. The difference in the palatal rugae pattern between two different ethnic groups may be factored on wide genetic and geographic separation. However, the role of genetics in rugae pattern needs to be evaluated.

REFERENCES

Acharya, A.B. & Sivapathasundaram B. 2007. Forensic odontology. In Rajendran R. & Sivapathasundaram B. (ed.), *Shafer's texbook of oral pathology*: 886. New Delhi: Elsevier Publications.

Bansode, Shriram C. & Meena M. Kulkarni. 2009. Importance of palatal rugae in individual identification. *J Forensic Dent Sci* 2(1): 77–81.

Caldas, M.T., Magalhães T. & Afonso A. 2007. Establishing identity using cheiloscopy and palatoscopy. *Forensic Sci Int* 165(1): 1–9.

Hauser, G., Daponte A. & Roberts M.J. 1989. Palatal rugae. *J Anat* 165: 237–249.

Kapali, Sunita, Grant Townsend, Lindsay Richards & Tracey Parish. 1997. Palatal rugae patterns in Australian Aborigines and Caucasians. *Aust Dent J* 42(2): 129–133.

Muthusubramanian, M., Limson K.S. & Julian R. 2005. Analysis of rugae in burn victims and cadavers to stimulate rugae identification in cases of incineration and decomposition. *J Forensic Odontostomatol* 23(1): 26–29.

Nayak, P., Acharya A.B., Padmini A.T. & Kaveri H. 2007. Differences in the palatal rugae shape in two populations of India. *Arch Oral Biol* 52(10): 977–982.

Paliwal, Aparna, Sangeeta Wanjani & Rajkumar Parwani. 2010. Palatal rugoscopy: Establishing identity. *J Forensic Dent Sci* 2(1): 27–31.

Segelinck, S.L. & Goldstein L. 2005. Forensic application of palatal rugae in dental identification. *Forensic Exam Spring* 14(1): 44–47.

Shetty, K.S., Kalia S., Patil K. & Mahima V.J. 2005. Palatal rugae in Mysorean and Tibetan populations. *Ind J Dent Res* 16(2): 51–55.

Effect of acrylamide in steeping robusta coffee (*Coffea canephora* var. *robusta*) on memory function and histopathological changes of brain cells in male rats (*Rattus norvegicus*)

D.Y. Lestari, M. Bahrudin, Rahayu, Fadhil & A.W. A'ini
Faculty of Medicine, Muhammadiyah Malang University, Jawa Timur, Indonesia

ABSTRACT: Coffee is known to contain acrylamide, a carcinogenic substance, that will lead to changes in the chemical structure of DNA, resulting in the degeneration of brain cells. It also contains caffeine, melanoidin, chlorogenic acid, and flavonoids that have been predicted to improve memory function. Objective: to evaluate the effect of steeping robusta coffee (*Coffea canephora* var. *robusta*) on memory function using the *Morris water maze* test and on histopathological changes of brain cells in male rats (*Rattus norvegicus*). Methods: this research is a true experiment with the posttest only control group design. The samples were divided into five groups. Group 1 received no treatment (normal control); groups II, III, IV, and V received steeping robusta coffee (*Coffea canephora* var. *robusta*) at different doses (0.36 ml, 0.72 ml, 1.44 ml, and 2.16 ml) for 28 days. Results: for memory function, the one-way ANOVA test showed significant effects (sig = 0.000, $p < 0.05$). Tukey's test showed a significant value of $p < 0.05$ in the groups treated with doses of 0.36 ml, 0.72 ml, 1.44 ml, and 2.16 ml, which differed significantly from the normal control group, but the group treated with the dose of 0.36 ml was not significantly different from the negative control group. Pearson's correlation test showed a value of 0.953, which means that when a higher dose of steeping robusta coffee is given, a higher time is required to find a *hidden platform*. For histopathological changes, the one-way ANOVA test showed $p < 0.05$, which means that there is a significant effect. Tukey's test showed a significant value of $p < 0.05$ in the treatment group. The strength of the relationship shows that steeping robusta coffee increases the number of edema cells in the brain (92.2%). Conclusion: steeping robusta coffee can cause a decrease in the memory function and histopathological changes of brain cells in male rats.

Keywords: Acrylamide, coffee robusta, memory function, brain cells

1 INTRODUCTION

Coffee is a drink that is very popular in the public because of its taste and flavor. Based on studies, coffee has been found to have acrylamide content and high organic acid content. Acrylamide is a known free radical in animals and in humans, and its specific doses can be toxic to the nervous system of animals and humans.[1] Acrylamide has the potential to exert neurotoxic effects, which affect catecholamine levels and several enzyme activities in the brain.[2] On the other hand, coffee also contains caffeine, which in small doses can improve the memory function.[3] On the other hand, coffee also contains chlorogenic acid, a highly antioxidant compound, that can break the chains of free radicals caused by acrylamide.

Coffee has both positive and negative effects depending on the content, and this makes the author to examine the effect of steeping coffee in graded doses on the memory function and histopathological changes of the brain cells in male Wistar rat strains (*Rattus norvegicus*).

2 METHODS

This research is a true experiment using the posttest only control group design. Changes in the memory function in male Wistar rat strains were assessed using the Morris water maze test, and the histopathological changes in the brain in male Wistar rats were also assessed.

This study used 25 healthy male rats (*Rattus norvegicus* Wistar strain) aged 10–12 weeks and weighing 180–250 grams. Rats were divided into five groups with each group containing a total of five rats. Group 1 (negative control) consisted of rats that were fed with BR-1 and were not given any treatment. Groups 2, 3, 4, and 5 are consisted of rats that were given steeping robusta coffee (*Coffea canephora* var. *robusta*) at different doses

(0.36 ml, 0.72 ml, 1.44 ml, 2.16 ml) for 28 days. Prior to the treatment on rats, the process of adaptation to water maze was conducted for 3 days and then the water maze was tested for 28 days to determine the memory function in each group of rats by observing the average time taken by a rat to find the hidden platform. After treatment, the rats were examined to assess the histopathological changes in the brain. Histopathological changes were assessed by the presence of glia cells that were edematous, characterized by the piknotik cell nucleus, and the vacuolated cytoplasm (halo). Observations were made in five visual fields at 400× magnification.

3 RESULTS

Based on the results, the difference in the median time from administration of varying doses of robusta coffee (*Coffea canephora* var. *robusta*) to the time when rats find a hidden platform. There is a tendency that the higher dosage of brewed robusta coffee (*Coffea canephora* var. *robusta*) will increase the average time needed to find the hidden platform. Except for group 2, there occurred a slight decrease in the average time from the normal group. From the amount of edema in the brain cells of rats, we obtained that the greater the dose of coffee, the more the number of glial cells that experienced edema.

Recapitulation data analysis one-way ANOVA test well to changes in memory function as well as the number of cell edema indicates that the value of sig = 0.000 ($p < 0.005$). It can be concluded that there is a significant relationship between administration of steeping robusta coffee and changes in memory and histopathological functions of the brain in male Wistar rats.

4 DISCUSSION

In this study, the assessment of memory function obtained different results with research done by Angelucci (2002) and Nehlig (2010), evidenced by data such as the average time required to find a hidden platform in the normal group and the treatment group. Factors that distinguish this study from previous research studies conducted by Angelucci (2002) and Nehlig (2010), where pure caffeine preparation was used, is the use of steeping robusta coffee (*Coffea canephora* var. *robusta*), because the Indonesian people consume more steeping coffee of the pure caffeine.[3,4]

Based on the statistical test to the average time needed to find a hidden platform that shows the memory function of Wistar rat strains during 28 days, it was found that the normal control group had the average number of times higher than the group treated with Award steeping robusta coffee (*Coffea canephora* var. *robusta*) 0.36 ml dose of 131–220 mg caffeine, although the statistics showed no significant results. This shows that the administration of steeping robusta coffee at a dose of 0.36 ml equivalent to one cup of coffee may improve the memory function.

In the normal group, the average number of times is lower than that in the treatment group with steeping Award robusta coffee (*Coffea canephora* var. *robusta*) at different doses (0.72 ml, 1.44 ml, 2.16 ml) containing caffeine (262–440 mg, 524–880 mg, 786–1320 mg). This indicates that administration of doses of robusta coffee steeping (0.72 ml, 1.44 ml, 2.16 ml) or its equivalent (2, 4, and 6 cups of coffee) can reduce the memory function of rats, determined by calculating the average number of times needed to find the hidden platform.

The results of this study have less in common with the results of Han et al (2007) on the effect of

Table 1. Effect of steeping robusta coffee (*Coffea canephora* var. *robusta*) on the average time required to find a hidden platform (seconds) as well as the average amount of cell edema.

Treatment	Average time required to find a hidden platform (seconds)	The mean number of cells edema (5 visual fields)
K1	65,5525	18
K2	64,5375	19.45
K3	66,7850	22.8
K4	69,5700	27.3
K5	71,8750	29

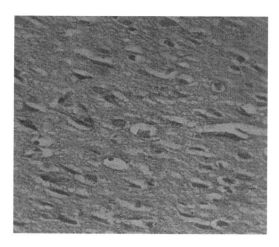

Figure 1. Microscopic images of brain cells induced by acrylamide with HE staining (400× objective). Red arrow indicates cell edema.

pure caffeine administration at a dose of 300 mg on the memory function of rats for 28 days. The study showed that there are significant differences in the mean time between the normal control group and the treatment groups. In the normal group, the average number of times is lower than that in the treatment group administered with a dose of 300 mg of pure caffeine. This shows that the administration of pure caffeine in high doses can reduce the memory function.[5]

Additionally, steeping robusta coffee contains acrylamide, which is the end product in the process of roasting coffee, appears as a clear crystal, is odorless, easily soluble in water, and present in many other organic solvents, but insoluble in benzene and heptane. A cup of coffee contains acrylamide at a concentration ranging from 0.32 to 1.46 g/30 ml in 6 g of coffee. Acrylamide is a harmful substance that has the potential to cause cancer in about 2% of cases each year in the world. Acrylamide is usually found in foods that are processed using high temperatures above 150°C. Acrylamide is not formed at a temperature below 120°C.[6]

Acrylamide is a compound containing ammonia, a compound that affects brain metabolism through increased synthesis of glutamine and ketoglutarate; these two ingredients affect Kreb's cycle, leading to the loss of ATP molecules. This resulted in impaired cell oxidation process. The release of calcium ions from this mitochondrial homeostasis results in disruption in the beginning of the death of the cell, so that it can reduce the memory function.

Acrylamide has the potential to exert neurotoxicity that results in the central and peripheral nervous system damage, autonomic, which in turn can cause fatigue, dizziness, sleepiness, and difficulty in remembering. The results of clinical trials show that acute exposure to high doses of acrylamide triggers the signs and symptoms of central nervous disorders. Exposure to smaller doses of acrylamide for a long term can trigger disturbances in the peripheral nervous system.[7]

The use of excessive coffee in high doses (> 500 ml), equivalent to two cups of coffee, can cause cardiovascular system disorders, such as increased heart rate and blood pressure. Hence, in circumstances where a person consumes too much coffee, there will be increased heart rate, which will result in anxiety, and therefore, it is difficult to learn and concentrate.[5]

This study is the beginning of the influence of robusta coffee (*Coffea canephora* var. *robusta*), where there are several limitations of the coffee used only method brewed with hot water instead of using pure caffeine or acrylamide purification extract and the dose is too large in this study. Therefore, further research should be conducted using pure caffeine or steeping robusta coffee with a lower dose.

REFERENCES

[1] WHO, 2002. Health Implication of Acrylamide. http://www.who.int/enity/foodsafety.
[2] Nehlig, Astrid. 2010. Is Caffeine a Cognitive Enhancer? Journal of Alzheimer's Disease 20 (2010) S85–S94 S85 DOI 10.3233/JAD-2010-091315 IOS.
[3] Angelucci ME, Cesario C, Hiroi RH, Rosalen PL, Da Cuncha C. 2002. Effect of caffeine on Learning And Memory in Rats Tasted In The Morris Water Maze. Braz J Med Biol Res. 35(10):1202–8.
[4] Han, Jin-Hee. Kushner, Steven A. Yiu, Adelaide P. Cole, Christy Joe. Matynia, Anna. et al. 2007. Neuronal Competition and Selection During Memory Formation. Science. Vol 316, Issue 5823. pp. 457–460.
[5] Harahap, Yahdiana. 2006. Pembuatan Akrilamida Dalam Makanan dan Analisisnya. Makalah Ilmu Kefarmasian, Vol III. No 9. pp.107–116.
[6] Tanseri, L. 2010. Pengaruh Susu Terhadap Kadar Acrylamida Dalam Kentang Goreng Simulasi: 34–35.

Effects of aerobic exercise and a high-carbohydrate diet on RBP4 expression in rat skeletal muscle

Nuroh Najmi
Faculty of Medicine, Universitas Padjadjaran, Bandung, Indonesia

Yunia Sribudiani
Department of Biochemistry and Molecular Biology, Faculty of Medicine, Universitas Padjadjaran, Bandung, Indonesia
Study Center of Medical Genetics, Faculty of Medicine, Universitas Padjadjaran, Bandung, Indonesia

Bethy S. Hernowo
Department of Pathology of Anatomy, Faculty of Medicine, Universitas Padjadjaran, Bandung, Indonesia
Dr. Hasan Sadikin Hospital, Bandung, Indonesia

Hanna Goenawan, Setiawan, Vita M. Tarawan & Ronny Lesmana
Physiology Division, Department of Anatomy, Physiology and Biology Cell, Faculty of Medicine, Universitas Padjadjaran, Bandung, Indonesia

ABSTRACT: Retinol-Binding Protein 4 (RBP4) is a vitamin A transport protein that plays a role in metabolism. The aim of this study was to analyze the effect of aerobic exercise and a high-carbohydrate diet on the immunoexpression of RBP4 in the skeletal muscle of rats. A total of 28 male rats were divided into four groups: normal diet, sedentary (K1); high-carbohydrate diet, sedentary (K2); normal diet, exercise (K3); and high-carbohydrate diet, exercise (K4). The rats in the exercise groups were subjected to running on a treadmill 5 days per week for 16 weeks with a speed of 15–23 m/min for 30 minutes. At the end of the study, RBP4 immunoexpression in the skeletal muscle of rats was examined. The results revealed that the RBP4 immunoexpression levels of all the groups were not significantly different. However, there was a weak correlation of body weight and exercise with the immunoexpression of RBP4.

Keywords: diet, aerobic exercise, RBP4

1 INTRODUCTION

Indonesia is one of the developing countries whose population growth rate increases rapidly. The increasing population growth is accompanied by increasing welfare needs. The increase in welfare needs changes the lifestyle. The Ministry of Health stated that the incidence of diabetes in Indonesia increases every year (RI 2013). Diabetes mellitus or diabetes type 2 is a metabolic disease caused by multi–factorial changes, including genetic and unhealthy lifestyles.

The main carbohydrate source in the Indonesian diet is rice. High carbohydrate intake can increase the risk factor of diabetes mellitus. Increased blood glucose can induce insulin release from pancreatic β-cells. Insulin mediates glucose uptake in peripheral tissues such as muscle, adipose tissue, and kidneys (Ma et al. 2016). The increasing glucose concentration can induce glycogen storage in the liver and muscle. High carbohydrate consumption for a long period increases the risk factor of insulin resistance (Alapatt et al. 2013). Retinol-Binding Protein 4 (RBP4) is a vitamin A transport protein, which is also one of the potential indicators of insulin resistance (Kotnik et al. 2011). RBP4 is a adipocytokine that is secreted mainly from the liver and adipose tissue. Some studies have revealed that serum RBP4 levels correlated with the metabolic syndrome. Therefore, RBP4 levels could be used to detect insulin resistance.

Exercise training has an important role in changing the metabolism activity in skeletal muscle. Exercise activates atypical Protein Kinase C (aPKC), which then increases glucose uptake in skeletal muscle. Another study has suggested that GLUT4 expression is increased 3–24 hours after exercise. Thus, regular exercise maintains high GLUT4 protein expression and subsequently improves the control of blood glucose levels.

The aim of this study was to analyze the effect of aerobic exercise and a high-carbohydrate diet on RBP4 expression in rat skeletal muscle (Kawaguchi R., Yu J., Honda J., et al. 2007).

2 MATERIALS AND METHODS

This study was approved by the ethical committee of the Universitas Padjadjaran Bandung.

2.1 Animals

A total of 28 male rats were divided into four groups: normal diet, sedentary (K1); high-carbohydrate diet, sedentary (K2); normal diet, exercise (K3); and a high-carbohydrate diet, exercise (K4). The rats were placed in individual cages with *ad libitum* access to food and water.

2.2 Diet

The high-carbohydrate diet consisted of a mixture of complex carbohydrates (70%) and simple carbohydrates (10%) (Table 1). The high-carbohydrate diet was given to the rats in the K2 and K4 groups for 16 weeks (Table 1).

Table 1. Diet composition.

	Weight (g)	Total weight (g)	Weight (%)	Calorie (kcal)
Normal diet				
Carbohydrate		600	60	2460
Simple	500			
Complex	100			
Protein		250	25	1025
Fat		150	15	1395
Water (ml)		100	100	
Vitamin (ml)		5		
Total				4880
High-carbohydrate diet				
Carbohydrate		800	80	3280
Simple	100			
Complex	700			
Protein		150	15	615
Fat		50	5	465
Water (ml)		100	100	100
Vitamin (ml)		5		
Total				4360

Explanation:
  for 1 kg diet
  1 g carbohydrate = 4.1 calorie
  1 g fat = 9.3 calorie
  1 g protein = 4.1 calorie
Source: Whitney.

2.3 Aerobic exercise

The rats were habituated for 5 days before subjected to physical exercise. Habituation of rats was measured by the Dishman Score. After the habituation period, the rats were subjected to the exercise intervention. They ran on a treadmill for 30 minutes/day at a speed of 15–23 m/min for 16 weeks. The rats were killed 24 hours after the last exercise. Gastrocnemius muscle was dissected and stored in buffer formalin solution.

2.4 Measurement

Rat body weight was measured every four weeks. Blood samples were taken from the periorbital sinus to measure blood glucose every 4 weeks.

2.5 Immunohistochemistry

At the end of the exercise period, the rats were killed by anesthesia and cervical dislocation. Gastrocnemius muscle was dissected and embedded in paraffin wax. The paraffin-embedded tissue blocks were cut using a microtome and processed for immunohistochemistry. After deparaffinization, the muscle was incubated with RBP4 primary antibody (BIOSS, US). After washing, the muscles were incubated with an appropriate secondary antibody. The slides were mounted, coverslipped, and examined under a light microscope.

3 DATA ANALYSIS

Data are expressed as means ± SD for each group. Using SPSS software, data were analyzed using two-way ANOVA followed by a *post hoc* test. The chi-square test was used to analyze immunohistochemistry.

4 RESULTS AND DISCUSSION

The results indicated that there were no significant differences in body weight between the groups (Table 2). The body weight of the K1 group (control) was the highest among the other groups.

Table 2. Average body weight.

Groups	Average body weight (gram) ± SD
K1	254,04 ± 56,42
K2	220 ± 32,92
K3	245,17 ± 57,53
K4	222,04 ± 37,26

However, the difference in body weight between the K1 and K3 groups (normal diet, exercise), was not significant. The rats in the high-carbohydrate diet group (K4) had a lower body weight than those in the normal diet group (K1).

The immunohistochemical analysis of RBP4 was carried out at the end of the study (Figure 1). The chi-square analysis showed that the histoscore of each group was not significantly different (Table 3). Nevertheless, there was a tendency towards a positive correlation between the expression of RBP4 and body weight.

This study shows that aerobic exercise can decrease RBP4 expression in the skeletal muscle of rats. Lim et al. (2008) explained that physical exercise can decrease RBP4 levels. However, the exact role of RBP4 is still unclear. Some studies have revealed that RBP4 is correlated with blood glucose levels, and other studies reported that RBP4 is correlated with body weight, obesity, and insulin resistance (Pullakhandam et al. 2012; Hoggard et al. 2012; Kotnik et al. 2011; Esteve et al. 2009). Janke et al. (2006) found that RBP4 expression is correlated with obesity and GLUT4 as has been observed from mRNA RBP4 in adipose tissue.

Under normal conditions, glucose entry into adipocytes is mediated by GLUT4, which suppresses the expression of RBP4. In animal studies, RBP4 expression has been shown to be positively correlated with GLUT4 expression in adipose tissue (Janke et al. 2006). Aerobic exercise significantly improves insulin resistance and decreases RBP4 levels in rats (Lim et al. 2008). It has been shown that RBP4 gene expression in the liver, adipose tissue, and muscle was higher in diabetic rats than in other groups (Sánchez et al. 2009). Chavez et al. (2009) reported that plasma RBP4 is elevated in diabetes mellitus type 2 and associated with impaired glucose tolerance but not associated with insulin resistance. Numao et al. (2012) reported that the levels of circulating RBP4 can be increased by exercise and may be associated with Triglyceride (TG) concentration in obese men.

5 CONCLUSION

Our data showed that there was no correlation between RBP4 expression in rat skeletal muscle and fasting blood glucose level. However, there was a weak positive correlation of body weight and exercise with RBP4 expression in the skeletal muscle of rats.

ACKNOWLEDGMENTS

We thank the Animal Laboratory Pharmacology Universitas Padjadjaran and Immunohistochemistry Laboratory Hasan Sadikin Hospital for their technical support. This study was supported by grant from DIKTI KLN to Ronny Lesmana (2014).

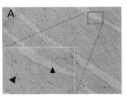
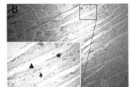
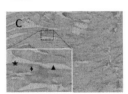

Figure 1. RBP4 expression in the skeletal muscle of rats. A: Gastrocnemius of the K3 group (negative); B: Gastrocnemius of the K2 group (positive weak); C: Gastrocnemius of the K3 group (medium weak). Each image was taken at a magnification of 100x. Description of the symbols: ↑ nucleus; ▲ cytoplasm; ★ positive RBP4 expression.

Table 3. Immunohistochemical score of RBP4 expression.

Groups	Average immunoexpression score ± SD
K1	2 ± 2
K2	1.167 ± 1.63
K3	1.33 ± 1.65
K4	0.83 ± 1.32

REFERENCES

Alapatt, P. et al., 2013. Liver retinol transporter and receptor for serum retinol-binding protein (RBP4). *Journal of Biological Chemistry*, 288(2): 1250–1265.

Chavez, A.O. et al., 2009. Retinol-binding protein 4 is associated with impaired glucose tolerance but not with whole body or hepatic insulin resistance in Mexican Americans. *Am J Physiol Endocrinol Metab*. 296: 758–764.

Esteve, E., Ricart, W. & Fernández-Real, J.M., 2009. Adipocytokines and insulin resistance: the possible role of lipocalin-2, retinol binding protein-4, and adiponectin. *Diabetes care*, 32 Suppl 2: S363–S367.

Hoggard, N. et al., 2012. Serum levels of RBP4 and adipose tissue levels of PTP1B are increased in obese men resident in northeast Scotland without associated changes in ER stress response genes. *International Journal of General Medicine*, 5: 403–411.

Janke, J. et al., 2006. Retinol-binding protein 4 in human obesity. *Diabetes*, 55(10): 2805–2810.

Kawaguchi R., Yu J., Honda J., et al. A membrane receptor for retinol binding protein mediates cellular uptake of vitamin. *Science*. 2007; 315(5813): 820–825.

Kotnik, P., Fischer-Posovszky, P. & Wabitsch, M., 2011. RBP4: A controversial adipokine. *European Journal of Endocrinology*, 165(5): 703–711.

Lim, S. et al., 2008. Insulin-sensitizing effects of exercise on adiponectin and retinol-binding protein-4 concentrations in young and middle-aged women. *Journal of Clinical Endocrinology and Metabolism*, 93(6): 2263–2268.

Ma, X. et al., 2016. RBP4 functions as a hepatokine in the regulation of glucose metabolism by the circadian clock in mice: 354–362.

Numao, S. et al., 2012. Effects of Exercise Training on Circulating Retinol-Binding Protein 4 and Cardiovascular Disease Risk Factors in Obese Men. *The European Journal of Obesity*: 845–855.

Pullakhandam, R. et al., 2012. Contrasting effects of type 2 and type 1 diabetes on plasma RBP4 levels: The significance of transthyretin. *IUBMB Life*, 64(12): 975–982.

Ri, B.P. dan P.K. kementerian K., 2013. *Riset Kesehatan Dasar [RISKESDAS]*.

Sánchez, O. et al., 2009. Skeletal muscle sorbitol levels in diabetic rats with and without insulin therapy and endurance exercise training. *Experimental diabetes research*, 2009: 737686.

Whitney EN RS. Understanding Nutrition. 12th ed. USA: Wadsworth Publishing; 2010.

… Advances in Biomolecular Medicine – Hofstra, Koibuchi & Fucharoen (Eds)
© 2017 Taylor & Francis Group, London, ISBN 978-1-138-63177-9

Correlation of physical activity and energy balance with physical fitness among the professors of the University of Padjadjaran

Juwita Ninda[1], Leonardo Lubis[2], Ambrosius Purba[1], Ieva Baniasih Akbar[1,11], Wahyu Karhiwikarta[1], Setiawan[1], Vita Murniati Tarawan[1], Reni Farenia[1], Gaga Irawan Nugraha[3], M. Rizky Akbar[4,12], Deni Kurniadi Sunjaya[5], Sylvia Rachmayati[6,12], Irma Ruslina[7,12], Hanna[1], Teddy Hidayat[8,12], Putri Tessa[1], Nova Sylviana[1], Ronny[1], Titing Nurhayati[1], Yuni Susanti Pratiwi[1], Novi Vicahyani Utami[9], Juliati[1], Siti Nur Fatimah[10], Yunia Indah[1] & Fathul Huda[1]

[1]*Physiology Division, Department of Anatomy, Physiology and Cell Biology, Universitas Padjadjaran, West Java, Indonesia*
[2]*Anatomy Division, Department of Anatomy, Physiology and Cell Biology, Universitas Padjadjaran, West Java, Indonesia*
[3]*Department of Biochemistry and Molecular Biology, Universitas Padjadjaran, West Java, Indonesia*
[4]*Department of Cardiology, Universitas Padjadjaran, West Java, Indonesia*
[5]*Department of Public Health, Universitas Padjadjaran, West Java, Indonesia*
[6]*Department of Clinical Pathology, Universitas Padjadjaran, West Java, Indonesia*
[7]*Department of Physical Medicine and Rehabilitation, Universitas Padjadjaran, West Java, Indonesia*
[8]*Department of Psychiatry, Universitas Padjadjaran, West Java, Indonesia*
[9]*Department of Clinical Pharmacology, Universitas Padjadjaran, West Java, Indonesia*
[10]*Nutrition Division, Department of Public Health, Faculty of Medicine, Universitas Padjadjaran, West Java, Indonesia*
[11]*Faculty of Medicine, Universitas Islam Bandung, West Java, Indonesia*
[12]*Hasan Sadikin General Hospital, West Java, Indonesia*

ABSTRACT: Background: Physical inactivity is the fourth leading cause of death in the world. It may lead to the development of degenerative diseases. Objective: to examine the correlation of physical activity and energy balance with physical fitness among the professors of Unpad. Methods: Professors of Unpad undertook a questionnaire from *ALG* (*Academic Leadership Grant*) for 1 month. A total of 142 elderly professors (females (n = 34) and males (n = 115)) were selected in this study. They were classified into four groups: pre-elderly (≥ 50–60 years), young elderly (≥ 60–70 years), old elderly (≥ 70–80 years), and very old elderly ≥ 80 years. The professors were administered a questionnaire on physical activity that is called the *GPAQ* (Global Physical Activity Questionnaire), which included 24-hour food intake recall, *TEE*, and a physical fitness test consisting of walking in 6 minutes. Results: Data were analyzed by descriptive statistics, *ANCOVA* analysis (p ≤ 0,05), and logistic regression. Conclusion: Physical activity with moderate intensity is effective for the elderly.

Keywords: Professor of Padjadjaran *ALG*, *GPAQ*, 24-hour Food recall, *TEE*, six-minute walking test

1 INTRODUCTION

1.1 Overview

The elderly are otherwise healthy adults who experience a number of changes in physiological processes. Based on a report by the Central Bureau of Statistics in 2016, the number of elderly aged 65 years and above in Indonesia amounted to 14,233,177 people. The fairly large number of elderly and the growing number of health care need to stay productive for a healthy life. In this regard, the government passed a law (article 138 of Law 36/2009), which makes mandatory that: the elderly must be nurtured and directed to stay healthy, fit, happy, prosperous, and productive, and be able to meet the needs so that it does not become a burden on other communities. The elderly were divided into pre-elderly .≥ 50–60 years old, young elderly .≥ 60–70 years, old elderly .≥ 70–80 years, and very old elderly .≥ 80 years. According to a report by the *WHO* and *MOH* in 2009, elderly people aged 65 years will represent 16% of the total population in 2050.[1,2]

The decline of function in physical condition is settled, and reflected as a physiological profile in the form of fitness criteria, including 6-minute walking test and elderly physical activity questionnaire *GPAQ* (Global Physical Activity Questionnaire). In addition, a decline in the physical condition can also be reflected by excessive or less food intake in the elderly. The decline in the physical condition of the elderly is closely related to an impaired balance between food intake and energy expenditure.[3]

The calculation of this questionnaire is done by counting the number of metabolic equivalent turnover/week/min/physical activity. Physical activity was observed in the form of work, leisure, travel, and sitting down. The study was conducted among the professors of the University of Padjadjaran with questionnaires *GPAQ* to measure the physical activity for 1 week.

In addition, research was also conducted on the correlation of physical activity with physical fitness test (6-minute walking), and the energy balance consisting of 24-hour food intake recall and the use of energy (total energy expenditure) with a physical fitness test of 6-minute walking, namely the ability to perform activity to the maximum and thus improve health and avoid periods of inactivity.[4,5,6]

1.2 GPAQ: The physical activity questionnaire in the elderly

Weekly physical activity in the elderly was measured using the *GPAQ*. Here are four areas of coverage of exposure to physical activity: activity on weekdays, namely physical activities that require more energy than that expended each day. Physical activity and exercise outside of work are leisure time activities, which are often interpreted differently by society and called inactivity/lazy. Various types of physical activity, among others by using the transport way, like cycling require a lot of energy.[6,7,8]

GPAQ was developed by the WHO for physical activity surveillance in many countries. It collects information on physical activity participation in three settings (or domain) as well as sedentary behavior, comprising 16 questions (P1-P16). The domains are: activity at work, travel to and from places, and recreational activities.

GPAQ MET was used in the calculation, as given in Table 1.[9]

The equation of total physical activity *MET*-minute/week is = (P2 × P3 × 8) + (P5 × P6 × 4) + (P8 × P9 × 4) + (P11 × P12 × 8) + (P1 4 × P15 × 4).[9]

1.3 Physical fitness in the elderly

A 6-minute walking test was conducted to measure the level of fitness in healthy people, by evaluat-

Table 1. Level of activity measurement and value-based GPAQ MET (metabolic equivalent turnover).

DOMAIN	MET
Work	– Moderate *MET* = 4 – Strenuous *MET* = 8
Transportation	Cycling and Walking *MET* = 4
Recreation	– Moderate *MET* = 4 – Strenuous *MET* = 8

Source: *GPAQ, WHO*.

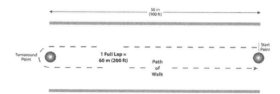

Forms walk test field for 6 minutes

Figure 1. The road test field for 6 m.
Source: *Insight PAH, Gilead Sciences, 2016*.

ing the entire system during exercise. The results of the physical fitness test of 6-minute walking were grouped into four categories: low, moderate, good, and excellent according to age group, namely ≥ 40–49 years, ≥ 50–59 years, ≥ 60–69 years, and ≥ 70–80 years (Figure 1).[10]

1.4 Food intake in the elderly

Food intake in the elderly must be considered to avoid excess and malnutrition. Many elderly people in Indonesia consume bread, bananas, chocolate, ice cream, instant noodles, sweet coffee along with cream, milk, and sweet tea that have a glycemic index, which is moderate to high, but not particularly a good nutritional value, and there was an increase in calories from adding creamer, milk, and sugar. Intake of all kinds of food and beverages by an individual is called a 24-hour recall method.

Food intake in the elderly is grouped according to the level of intensity in kcal/day and the category includes intake on sedentary activity, the activity of low intensity and moderate-intensity activities by weight based on the three age groups, namely 31–50 years, 51–70 years and .≥ 71 years.[11]

1.5 Energy expenditure in the elderly

1.5.1 BMR in the elderly
BMR is the minimum caloric requirement needed to survive when the condition of the body is at rest. *BMR* calculation is done using the formula of Harris Benedict.

Based on height, weight, gender, and age, *BMR* describes 45–70% of total daily energy expenditure by the Harris Benedict formula:

Men = 66 + (13.7 × Weight) + (5 × Height) − (6.8 × age),
Women = 65.5 + (9.6 × Weight) + (1.8 × Height) − (4.7 × age).

1.5.2 Total Energy Expenditure in the elderly

Total Energy Expenditure (*TEE*) is defined as *RMR* plus energy consumed during activity (*AEE*) and Diet-Induced Thermogenesis (*DIT*). *TEE* component consists of *RMR*, metabolism during activity, and metabolism of food. The formula to calculate the total energy expenditure, among others: *TEE* = *BMR* × factor stress factor X activity.[12,13]

2 METHODS

The University of Padjadjaran consists of 142 people given *ALG* (Academic Leadership Grant) for 1 month. Professors of Unpad have fulfilled the inclusion and exclusion criteria and gave informed consent. A total of 34 females and 115 males, undertook the questionnaires *GPAQ*, 24-hour food intake, and physical fitness measuring 6-minute walking test at different elderly groups, namely pre-elderly ≥ 50–60 years (n = 25), young elderly ≥ 60–70 years (n = 52), old elderly ≥ 70–80 years (n = 55), and very old elderly ≥ 80 years (n = 10).

The data are then processed to a unit of measurement *GPAQ MET* (*Metabolic Equivalent Turnover*)/min/week. Then, it is converted into units of calories.

Measurement of food intake by 24-hour food recall calories is done by interviewing the amount of food for 24 hours, and the professors were asked to remember the kind of food that is consumed on the previous day.

TEE measurement is a measurement of energy balance. Before obtaining *TEE*, *BMR* calculation is done in advance using the formula of Harris Benedict. Physical fitness measurement by 6-minute walk test was conducted by professors with a fast walk or his/her best, and the distance traveled by a professor was recorded by a team of enumerators *ALG* in meters and then put in the category of physical fitness.[14,15]

This study used *ANCOVA* to determine the relationship between variables, namely physical activity and physical fitness, and energy balance and physical fitness in different age groups of professors of the University of Padjadjaran.[16]

The relationship between physical activity and energy balance with physical fitness was determined using linear regression. In addition, the minimum value of *OR* was significantly set at 2.[17]

3 RESULTS

ALG questionnaire measurements have been conducted by 142 professors of Unpad, grouped according to the age group of the elderly. Measurement of physical activity, namely *GPAQ* in the elderly is measured by *MET*/min/week and activity energy expenditure (Table 2). Measurements showed professor by the *MET* did not reach 50% and category light. While based on the activity energy expenditure professor reached above 50% in light category. *ANCOVA* analysis is used in the measurement of physical activity *MET*/minutes/week, which has a closer relationship than the other independent variables with physical fitness. Based on logistic regression with regard to confounding factors, *TEE* is significantly more than other variables, namely *OR* .≥ 2.[5,6,9,15]

In Table 3, based on the food intake of more than 50% of professors, they were categorized as fair and *ANCOVA* analysis p value is less significant than other variables. Based on logistic regression, food intake is closely linked to physical activity. If the energy expenditure of professors was more than 50%, they were in the moderate category and have meaningful *ANCOVA* analysis p ≤ 0.05, as well as logistic regression p ≤ 0.001. The results of the test measurements of 6 minutes more than 50% are in the poor category. In Figure 2, the total value of the *TEE*, and the sum of the activity energy expenditure, resting metabolic rate, and thermal effect of food are shown. The meas-

Table 2. Percentage between physical activity *MET* and *AEE*, and *ANCOVA* analysis between physical activity and physical fitness of 6-minute walking test. P significant ≤ 0,05.

Category	Very light	Light	Moderate	Heavy	Strenuous	P	CI 95%	
MET GPAQ	24%	32%	18%	20%	6%	0,26	313	111,067
Energy activity	0%	57%	43%	0%	0%	0,76	61,803	81,894
With food intake						0,32	182,953	187,970
With TEE						0,35	1083,634	715,621

Table 3. Percentage between food intake, *TEE*, six-minute walking test, and *ANCOVA* analysis between food intake, *TEE* test to physical fitness. *P* significant at ≤ 0,05.

Category	Poor	Fair	Moderate	Good	Very good	P	CI 95%	
Food intake	0%	76%	24%	0%	0%	0,9	87,313	69,153
TEE	0%	28%	72%	0%	0%	0,05	140,18	−7,605
SMWT	63%	27%	7%	3%	0%	0,048		

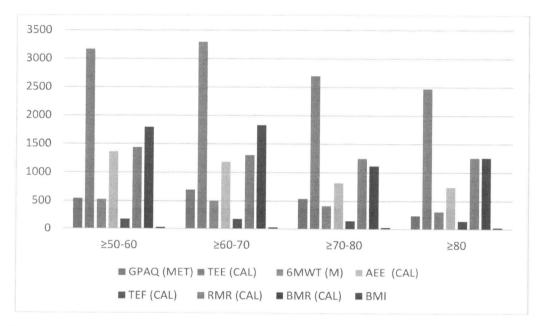

Figure 2. Graph of total energy expenditure and *GPAQ* with six-minute walking test.

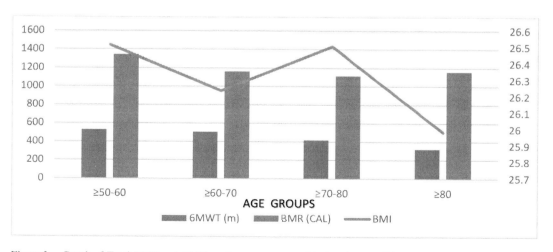

Figure 3. Graph of Total *BMR* and *BMI* based on age groups with six-minute walking test.

Table 4. Regression between physical activity to food intake and *TEE*. Regression between six-minute walking test to *AEE*. P significant at ≤ 0,05. 95% *CI* significant > 2.

Logistic regression	P	OR	CI 95%	
Food intake	0,001	1,6	0,197	1,26
Energy activity	0,05	1,96	0,7	3,69
TEE	0,01	0,52	1,145	7,248
6-minute walking test	0,823	1,06	0,629	1,790

urement results of *TEE, AEE, TEF, RMR, GPAQ*, and 6-minute walking test decreased successively in the age group of .≥ 80 years, .≥ 70–80 years, .≥ 60–70 years and .≥ 50–60 years. This was caused by a decrease in *BMR* and *BMI* in Figure 3.[12]

4 DISCUSSION

The results in Table 2 show that the *MET* are in the category of light 32%, very light 24%, heavy 20%, moderate 18%, and strenuous 6%. *MET*/min/week of professors are in the light category, which means that the professors are more active, 32% lighter (activity energy expenditure in the light category). In this category of physical activity, it has a value of ≤3 *METS*. *MET* is the unit of measurement in the measurement of the absolute intensity of physical activity, i.e., shopping, running to the office, sitting in front of a computer, clean bed, eating, preparing food, and washing dishes. A half of them were in the moderate category, with a *MET* value .≥ 4. Activities of moderate intensity showed maximal oxygen consumption in a variety of activities, e.g., sweeping the floor, walking briskly, dancing, dusting house, wiping the windows, and playing basketball. Vigorous-intensity activity has a *MET* value of .≥ 8, and the activities include running, swimming, playing ball, jumping rope, and carrying heavy objects. While the activity energy expenditure describes the amount of energy during exercise and is a determinant of metabolism in the overall activities carried out day-to-good either planned or not. The better the value of activity energy expenditure was getting better at preventing weight gain and obesity.[5,14,15]

The results in Table 3 show the food intake of professors of Padjadjaran in the category quite 76% and was 24% with a value of $p = 0.9$. Food intake has less correlation when compared with *TEE* variables, namely moderate category 72% and sufficient 28%, $p ≤ 0.05$. The 6-minute walking test has a close correlation with the intake of food and *TEE*, namely $p ≤ 0.048$ in the following categories: poor 63%, fair 27%, good 7%, and very good 3%. The 24-hour recall method is described as obtaining data about the number of calories (energy) to record the type and amount of food consumed in periods of 24 hours the previous day.

With the degenerative process, the needs of most nutrition for the elderly are reduced, but the needs of most nutrients increase. The nutritional requirements are affected by the range of basal metabolism, which decreases with age accompanied by a decrease in energy needs per kilogram of body weight, resulting in weight loss of lean body mass. In the measurement of resting metabolic rate needed to maintain energy homeostasis by 60–80% *TEE* calculations. The other components are: activity energy expenditure of about 15% per day. While the Diet-Induced Thermogenesis (*DIT*) is 10% of the sum of *REE* + *AEE*, A decrease in total energy expenditure at the age of 60–90 years has decreased by 675 kcal in men and 459 kcal in women. In men, the reduction is 10% per year and it is 8% for women after the age of 60 years. Pursuant p-value *ANCOVA* analysis significantly results in value *MET GPAQ* $p = 0.26$, food intake and *TEE* $p = 0.35$ and *AEE* $p = 0.78$ with physical fitness. The results of the analysis of physical activity with food intake $p = 0.9$ and *TEE* significant $p ≤ 0.05$. While based on the 6-minute test, the significance is $p ≤ 0.05$. This means that the *TEE* and the 6-minute test have a closer relationship than food intake in the category fair.

In Table 4, based on logistic regression food intake, *TEE, AEE* has a positive regression, $p ≤ 0.001$ and $p ≤ 0.05$. The 6-minute walking test has less regression with physical activity. This means that the variable is vital to physical activity food intake and *TEE* with *CI* 95% .≥ 2 on the *TEE* and the *AEE*. Physical activity plays a role in energy balance, the *TEE* and the intake of food. The 6-minute test has a predictive role in physical activity. Figure 2 shows a graph of total energy expenditure at different age groups. The graph is based on components of energy expenditure, namely *RMR* plus energy consumed during activity (*AEE*) and Diet-Induced Thermogenesis (*DIT*). *TEE* component consists of *RMR*, metabolism during activity, and metabolism of food. The graph is divided into four age ranges as follows: .≥ 50–60 years, ≥ 60–70 years, ≥ 70–80 years, and ≥ 80 years.

The graph shows the average *TEE, GPAQ,* and 6-minute walking test per age group. Basal metabolic rate is decreased by 1–2% per year (400 kJ/day began to decline predicted by the age of 20–70 years with an increasing age), and it is a major component of energy expenditure and consists of 50–70% of total energy expenditure. The aging process correlates with food intake and expenditure as well as the balance of energy intake arrangement and also expenditure. Elderly disruption balance food intake, due to delay absorption of macronutrients by a decrease in the ability of the sense of smell, taste, hormones, and mediators metabolism. Impaired balance of food intake either overeating or eating less will affect the increase and decrease in energy expenditure and metabolic range.[12]

5 CONCLUSION

Overall, correlations of physical activity and energy balance with the physical fitness test 6-minute walking test have been obtained through research by *ALG*. According to the research, there is a correlation between physical activity and physical fitness among the different groups of elderly professors of the University of Padjadjaran. There is a correlation between physical activity and food intake among the different groups of elderly professors of the University of Padjadjaran. There is a correlation between physical activity and energy expenditure among the different groups of elderly professors of the University of Padjadjaran. The factor that largely contributes to physical fitness is physical activity. Moreover, physical activity with moderate intensity is effective for the elderly. Food intake and *TEE* have a role in physical activity. The results of the study indicate that professors of the University of Padjadjaran are expected to increase regular physical activity, programmed, and planned, so as to increase to a maximum category. In this study, there are limitations, namely lack of compliance in the 6-minute test run. We recommend that research results can be combined with cognitive function to a larger scale in various age groups.[15]

REFERENCES

[1] Abikusno, N. Turana, Y. Santika, A. 2013. *Gambaran Kesehatan:Lanjut Usia di Indonesia*. Jakarta: Buletin jendela data dan informasi kesehatan semester I.
[2] Topatimasang, R. 2013. *Memanusiakan Lanjut Usia: Penuaan Penduduk dan Pembangunan di Indonesia*. Yogyakarta: Surveymeter: INSISTPress.
[3] Jansen, M. Prins, R. Etman, A. Van Der Ploeg. Vries, S. Van Lenthe, F. 2015. *Physical Activity in Non Frail and Frail Older Adults*. Plos one 10(4):e0123168.
[4] Sumintarsih. 2006. *Kebugaran Jasmani untuk Lanjut Usia*. Yogyakarta: UPN Veteran 149–152.
[5] Purba, A. 2012. *Upaya pengendalian Proses Menua*. Bandung; Perhimpunan Ahli Ilmu Faal Olahraga Indonesia: 1–20.
[6] Chodzko, W. Proctor, D. Flatarone, M, Minson, C. Nigg, C. Salem, G. Et all. 2009. *Exercise and Physical Activity for older adults*. American College of Sports Medicine DOI:10.1249/MSS.0b013e3181a) c95c:1510–1530.
[7] Junaidi, S. 2011. *Pembinaan Fisik Lansia Melalui Aktivitas Olahraga Jalan Kaki*. Jurnal Media Ilmu Keolahragaan Indonesia Vol 1. Edisi 1 Juli; 17–21.
[8] Laksim, R. 2008. *Aktivitas Fisik Pada Lanjut Usia*. Universitas Negeri Yogyakarta: 3–4.
[9] Chan, M. *Global Physical Activity Questionnaire (GPAQ) Analysis Guide*. Geneva Switzerland: Prevention Of Noncommunicable Disease Department World Health Organization.
[10] Gozal, D. 2002. *ATS Statement: Guidelines For The Six-Minute Walk test*. American Thoracic Society Vol 166: 111–117.
[11] Triwijayanti, E. 2016. *Food Recall 24 Hours*. Wordpress: 1–10.
[12] Manini, T. 2011. *Energy Expenditure and Aging*. Ageing Res Rev 9(1): 1–26.
[13] Bompa, T. 2009. *Periodization: Theory and Methodology of Training, 5th ed*; Champaign: human kinetics.
[14] Guyton, H. 2006. *Textbook of Medical Physiology 11th ed*; Philadelphia: Elsevier Saunders.
[15] Ambrosius, P. 2016. *Kardiovaskular dan Faal Olahraga*. Bandung: Fakultas Kedokteran Universitas Padjadjaran.
[16] Dahlan, S. 2012. *Analisis Multivariat Regresi Logistik disertai Praktik dengan SPSS*. Epidemiologi Indonesia; Jakarta.
[17] Dahlan, S. 212. *GLM: Ancova & Repeated Measured Teori dan Praktik dengan SPSS*. Epidemiologi Indonesia: Jakarta.

Molecular biology of irreversible pulpitis: A case report

D. Prisinda & A. Muryani
Faculty of Dentistry, Universitas Padjadjaran, Bandung, West Java, Indonesia

ABSTRACT: Objective: Molecular biology provides opportunities to develop new strategies or agents for the treatment of a wide variety of diseases. The overall response of the tooth to injury represents the complex interplay between injury, defense, and regenerative processes. Methods: One-visit endodontic treatment was initiated on tooth 11 with irreversible pulpitis due to trauma. Result: Direct damage caused by injury or via exotoxin produced by microbes leads to the release of mediators that increase vascular permeability. Growth factors play a pivotal role in signaling the events of tissue formation and repair in the dentin–pulp complex. In our case study, the patient presented with irreversible pulpitis and no complaint was made after the treatment. Conclusion: Before the selection of treatment, it is important to know about the molecular biology of the diagnosis. Therefore, one-visit endodontic treatment on tooth 11 with a proper diagnosis can give the best result.

Keywords: molecular biology, irreversible pulpitis, acute inflammatory, one-visit endodontic

1 INTRODUCTION

The ideal number of visits for root canal treatment is still being debated. A Japanese study has shown that the result of successful one-visit endodontic treatment is 86% after 40 months. While Sjogren showed an opposite result, the success rate of which is small (approximately 68%).[1,2]

One-visit endodontic treatment is not done on the teeth with the case of the sinus tract, anatomical anomalies, acute apical periodontitis, and repeated root canal treatment.[2,4] Cases recommended for one-visit endodontic treatment are those that require a fast treatment, non-acute teeth, a simple anatomical shape of the root canal, and that the pulp disease be not spread to the periapical area,[2,4] for example in the case of trauma that results in an anterior tooth crown fracture involving the pulp as in our case. This case report will discuss one-visit root canal treatments on the first maxillary right incisor.

2 CASE REPORT

A male patient aged 24 years old visited the PPDGS clinic, which was part of Operative Dentistry Department RSGM FKG UNPAD, on December 13, 2011, with a complaint of fracture on the right upper anterior tooth and pain four days earlier. Pain was severe when the tooth was exposed to cold air, and when drinking cold water or eating sweet food. The patient took painkillers and the pain subsided. Nevertheless, he wanted his tooth to be treated due to pain and compromised appearance.

The results of the objective examination showed a fracture on tooth 11 with 1/3 left incisor crown and exposed pulp (Figure 1). Cold tests and Electric Pulp Tester (EPT) showed positive responses. Examination of percussion and the press showed negative responses, and the palpation test was also negative. Mobility examination showed a negative response. A good oral hygiene was maintained.

Radiographic examination showed an intact lamina dura, with no abnormalities in the periapical region (Figure 2). The patient was diagnosed

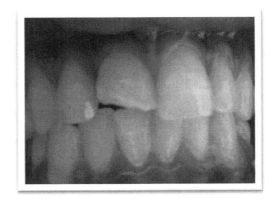

Figure 1. Clinical profile of a 1/3 incisor fracture on tooth 11.

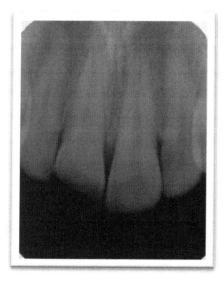

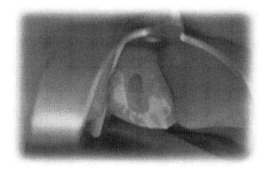

Figure 3. Tooth isolation with a rubber dam, cavity access opening.

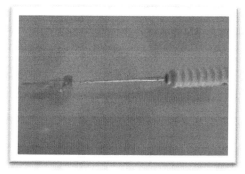

Figure 2. Photograph of early diagnosis.

with irreversible pulpitis tooth 11 with crown fracture. Prognosis was good. The treatment plan was one-visit treatment and follow-up with fiber post and all porcelain crowns.

3 CASE MANAGEMENT

Root canal treatment on tooth 11 begins with aseptic actions in advance by rubbing the povidone iodine solution on the labial and palatal sides. Thereafter, anesthesia infiltration is injected into the labial and palatal areas with an anesthetic solution of 0.5 cc on each side. Once the tooth is anesthetized and gingiva is pale, then isolation can be put on tooth 11 by using a rubber dam (Figure 3a), and then access is made on the palatal side of tooth 11 using a small round. Widening orifice is done using Z- and X-Gates endodontic burs (Dentsply Maillefer, Switzerland) (Figure 3b).

After properly opened, the pulp chamber is cleaned, extirpation of the pulp tissue is done inside the root canal, irrigation with 2.5% NaOCl (sodium hypochlorite), and then the canal is dried using paper points. (Figure 4) The working length is determined by the initial file no. 15 (K-files, Dentsply) using apex locator (Propex, Dentsply), with the result being 22 mm. Then, the root canal preparation is performed with a crown down technique according to working length using a ProTaper rotary needle (ProTaper, Dentsply) and 15% EDTA (Glyde, Dentsply, DeTrey, Switzerland) as a lubricant gel. Root canal preparation begins with a needle size of Sx, S1, S2, and F1–F5. Irrigation is done on each needle permutation with

Figure 4. Pulp tissue extirpation using an extirpation needle.

2.5% NaOCl. After the last irrigation using 2.5% NaOCl, a root canal is dried using paper points.

Photo trial is done according to the working length using gutta percha #F5. Radiographic result reveals the length of gutta percha in accordance with the working length (Figure 5). Root canals are filled with endomethasone sealer using a lentulo needle in the counterclock direction driven by a low-speed contra angle and gutta percha material filler #F5 as well as additional gutta percha # 15. The low-speed contra angle in the opposite direction (clockwise) and fillers gutta percha F5 numbers and additional gutta percha # 15 and # 20 are inserted into the root canal.

Filling is first done with the lateral condensation technique and then a radiographic image is taken to see the filling result. The radiographic image illustrates hermetic filling and is in accordance with the working length. Then, reduction of gutta percha as much as 2 mm below the orifice is done using a heated sterile excavator, and glass ionomer cement is put on top of it (GC, Gold-Label). It is then covered with a temporary restoration (Figure 6). The patient was asked to return 1 week later to do photo controls.

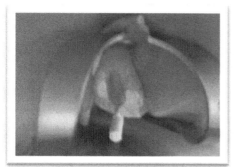

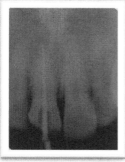

Figures 5a,b. Clinical and radiographic trial photograph.

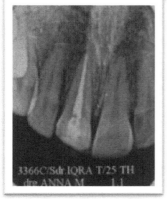

Figures 6a,b. Photograph of clinical and radiographic filling.

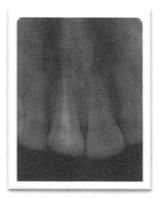

Figure 7. One week control after filling.

Control is done 1 week after filling. There was no complaint of pain. Percussion and pressure test are negative. The palpation test is negative and tooth mobility test is also negative. From the radiographic examination, hermetic filling is seen and there are no abnormalities in the periapical area (Figure 7).

Follow-up is planned for fiberpost preparation, fiberpost installation, and all porcelain crown making. The patient returns 19 days after installation. No complaints are addressed. Percussion and pressure tests give a negative result. Incisal contact under occlusion and functions shows no abnormalities. Crown adaptation to the tooth is good. No abnormalities are found on the gums. The patient had no complaints and is satisfied with the final result.

4 DISCUSSION

One-visit endodontic treatment has several advantages, namely the shorter treatment time, preferred by the patient, reducing the risk of flare ups and can prevent recontamination of the root canal. However, one visit treatment has disadvantages and limitations as follows, (1) long treatment time, especially affecting patients with TMJ disorders, (2) In the event of a flare-up, it cannot be drained easily, (3) In tough cases such as multiple or calcified roots, one visit treatment cannot be done.[1-5]

In one-visit endodontic treatment, there is no active infection after the contamination associated with the exposure of the pulp to the oral cavity, as the clinicians often extirpated the pulp, enlarged the canal, and sealed it in one visit. The key issue is thorough debridement of the root canal. If the patient is young and the tooth has minimal amounts of restoration, access to the canal system and complete debridement can be accomplished and then the root canals may be obturated. There are many additional factors that determine whether the canals can be sealed in a single visit, such as tooth anatomy, presence of calcification in the canal, access to the tooth and its canals, patient management, and degree of discomfort.[6]

The selection of one-visit root canal treatment on tooth 11 with a diagnosis of irreversible pulpitis is based on the absence of a complaint from the patient, the vitality of the tooth, the absence of periapical abnormalities, and the involved tooth with a single root and single root canal. All these reasons lead to the selection of one-visit treatment, therefore providing a good prognosis. Another thing that underlies the selection of this treatment is a good oral hygiene as well as the patient's desire for the treatment to be completed immediately.[6-7]

The acute inflammatory is a response to cell injury and is a sequence of events that is similar following injury caused by invading microbes or by physical mechanisms. Thus, direct damage by injury or via exotoxin produced by microbes leads to the release of mediators such as prostaglandins and leukotrienes that increase vascular permeability. The pivotal cell in an acute inflammatory response is the mast cell. This can be activated to release inflammatory cytokines such as TNF-α by binding microbes through pattern receptors, or to release inflammatory mediators such as histamine is an explosive exocytosis when the activation is via complement components that are a result of activation by microbes to the three pathways. IgE/allergen complexes and neuropeptides can also activate mast cells to release pharmacological mediators.

Microbial endotoxins activate macrophages to produce IL-1 and TNF-α, which have vasodilatory properties. The outcome of the abundant local mediator release is to loosen endothelial tight junctions and increase the adhesion of PMNs and monocytes and their migration into the surrounding tissues with which they come into contact and are able to phagocyte microbes. Fluid containing fibrinogen and antibodies is released into the area from the bloodstream, and thus edema protects the damaged area during repair.[6-7]

Invasion of bacteria or bacteria-derived factors into the pulp is a major etiologic factor in the inflammation of pulp tissues. This invasion can be initiated by, for example, caries or tooth fracture as shown in this case report (Trowbridge 2002). Pulpal nerve fibers react to inflammation by sprouting terminal branches and by altering their production of neuropeptides such as Calsitonin Gene-Related Peptide (CGRP) (Byers et al. 2003). Sensory nerve fibers play an important role in promoting the infiltration of immunocompetent cells into the pulp tissue. The initial infiltrating inflammatory cells are lymphocytes, macrophages, and plasma cells with a small amount of neutrophils.[7-9]

Acute pulpal inflammation, with the migration of abundant amounts of neutrophils toward the lesion, starts only when bacteria invade the reparative dentin formed under the lesion (Trowbridge 2002). The acute inflammatory response largely consists of a vascular reaction. Mediators released by various cells (e.g. serotonin, histamine, and neuropeptides such as substance P, CGRP, neurokinin A, and somatostatin) lead to alterations in pulp blood flow and increase the capillary permeability. Thus, plasma proteins and neutrophils gain better access to the inflamed area and can neutralize or phagocytose the irritant (Trowbridge 2002). Different types of T-lymphocytes (CD4+ helper, CD8+ cytotoxic), macrophages, neutrophils, dendritic cells, and plasma cells can be found in the inflamed pulp tissue, and their numbers increase with the progression of the disease (Jontell et al. 1998). Dendritic cells and also CD3+ (memory) T-lymphocytes accumulate under the lesion, while the accumulation of other immunocompetent cells is minor, suggesting that dendritic cells and memory T cells are critical in triggering immunological reactions of the pulp (Jontell et al. 1998). Severe pulpal inflammation may lead to an increased interstitial tissue pressure, which in turn will cause necrosis, since the pulp is enclosed inside rigid mineralized structures and thus it indicates a non-compliance environment.[8-10]

In this case study, root canal preparation is done with the crown down technique because it can remove more debris and microorganisms toward the coronal and maintain apical constriction, which can prevent the debris from being pushed to the apical region. The irrigating solution used is 2.5% NaOCl because it has an anti-microbial effect and can break the chains of proteins and damage bacterial DNA synthesis activity, as well as enhance lubrication. The use of 15% EDTA solution is effective in eliminating the smear layer, particularly inorganic components.[11,12]

In this case study, a lateral condensation root canal filling procedure was adopted using last number (F5) gutta percha with an additional small gutta percha #15 to provide hermetic filling. The filling is performed using gutta percha and endomethasone sealer containing corticosteroid that reduces pain and paraformaldehyde that inhibits bacterial growth and invasion.[13-14]

5 CONCLUSION

Before the selection of treatment, it is important to know about the molecular biology of the diagnosis. Therefore, in our case study, the selection of one-visit root canal treatment on tooth 11 with a diagnosis of irreversible pulpitis is based on the absence of a complaint from the patient, the vitality of the tooth, the absence of periapical abnormalities, and the involved tooth with a simple root canal, and this treatment can give the best result.

REFERENCES

[1] Siquiera, J.H. 2007. Periradicular repair after two visit endodontic treatment using two different intracanal medication compared to single visit endodontic treatment. *Braz Dent J* 18(4): 299–304.
[2] Stephen, C & Hargreaves, K.M. 2011. *Pathways of the pulp*. 10th ed. St. Louis: Mosby Elsevier.
[3] John, I.I & Bakland L. 2008. *Endodontics* 6th ed. Ontario: BC Decker Inc.
[4] Stephen, C & Burns R.C. 2002. *Pathways of the Pulp* 8th ed. United State of America: Mosby.
[5] Grossman, L.I & Oliet, S. 1995. *Ilmu Endodontik dalam Praktek*, (terj), 11th ed. Jakarta: EGC.
[6] Burtan, R & Louis, R. 2006. Endodontic microbiology. In Richard J.L, Robert A.B, Marylin S.L, Donald J.L (Ed). *Oral Microbiology and Immunology*. Washington DC: ASM Press.
[7] Peter, M.L & Michael, F.C. 2006. The immune system and host defense. In Richard J.L, Robert A.B, Marylin S.L, Donald J.L (Ed). *Oral Microbiology and Immunology*. Washington DC: ASM Press.
[8] Trowbridge, H.O. 2002. Histology of pulpa inflammation. In Hargreaves K.M & Goodis H.E (Ed). *Seltzer and Bender's Dental Pulp*. Illinois: Quintessence Publishing Co.
[9] Byers, M.R, Suzuki, H & Maeda, T. 2003. Dental neuroplasticity, neuro-pulpal interactions, and nerve regeneration. *Microsc Res Tech* 60(5): 503–515.
[10] Jontell, M, Okiji, T, Dahlgren, U & Bergenholtz, G. 1998. Immune defense mechanisms of the dental pulp. *Crit Rev Oral Biol Med* 9(2): 179–200.
[11] Mardewi, S.A. 2003. Endodontologi. Kumpulan naskah. Jakarta: Hafizh.
[12] Pekruhn, R.B. The incidence of failure following single visit endodontic therapy. 1986. *Journal of Endodontic* 2(2): 68–72.
[13] Torabinajed, M & Walton, R. Principles and practice of endodontics.1998. In Narlan Sumawinata, Prinsip dan praktik ilmu endodonsi edisi ke 2. Jakarta: EGC.
[14] Bergenholtz, G & Bindslev. 2003. Textbook of endodontology. Australia: Blackwell Munksgaard.
[15] Ford, P. 2004. Endodontics in clinical practice Harty's 4th ed. London: Wright.
[16] Inan, U. 2009. In vitro evaluation of matched taper single cone obturation with a fluid filtration method. *JCDA* 75(2): March.

Effect of different doses of X-ray irradiation on survival of human esophageal cells

I.M. Puspitasari & R. Abdulah
Department of Pharmacology and Clinical Pharmacy, Faculty of Pharmacy, Universitas Padjadjaran, West Java, Indonesia

M.R.A.A. Syamsunarno
Faculty of Medicine, Universitas Padjadjaran, West Java, Indonesia

H. Koyama
Department of Public Health, Gunma University Graduate School of Medicine, Maebashi, Japan

ABSTRACT: Radiotherapy or radiation therapy is one of the most effective therapies for cancer. Different doses of radiation are needed to kill different types of cancer cells. The aim of this study was to investigate the application of different doses of X-ray irradiation on noncancerous (CHEK-1 cells) and cancerous human esophageal cells (TE-8 cells). CHEK-1 cells and TE-8 cells were cultured using RPMI medium. Different doses of X-ray irradiation (0–8Gy) were applied. The clonogenic assay was performed to assess the survival of both cells. Plating efficiency (PE) and surviving fraction (SF) were calculated. SF between the control and irradiated groups was statistically significantly different ($P < 0.05$). SF between the non-cancerous and cancerous cells was not statistically significantly different ($P > 0.05$), except for 2Gy X-ray irradiation ($P < 0.05$). X-ray irradiation starting with a dose of 2Gy can decrease the survival rate of both noncancerous and cancerous cells. Therefore, 2Gy can be used as an irradiation dose for further study.

1 INTRODUCTION

Radiotherapy or radiation therapy is one of the most effective therapies for cancer (Puspitasari et al., 2014). The types of radiation used for cancer treatment are X-rays, gamma rays, and charged particles (National Cancer Institute). Radiation doses for cancer treatment are measured in a unit called gray (Gy), which is a measure of the amount of radiation energy absorbed by 1 kg of human tissue (National Cancer Institute). Different doses of radiation are needed to kill different types of cancer cells.

The clonogenic cell survival assay determines the ability of a cell to proliferate indefinitely, thereby retaining its reproductive ability to form a large colony or a clone. This cell is then said to be clonogenic (Anupama Munshi, 2005).

Clonogenic cell survival is a tool that was explained in the 1950s for the study of radiation effects (Anupama Munshi, 2005). Although clonogenic cell survival assays were initially used to study the effects of radiation on cells and have played an essential role in radiobiology, they are now widely used to determine the effects of potential agents for clinical applications (Anupama Munshi, 2005).

The aim of this study was to investigate the effect of different doses of X-ray irradiation on noncancerous (CHEK-1) cells and cancerous human esophageal (TE-8) cells.

2 METHODS

2.1 *Cell culture*

The CHEK-1 cell line was kindly provided by Dr H. Matsubara (Abdulah et al., 2009) and TE-8 cell line was obtained from RIKEN BRC, Ibaraki, Japan. Both cell lines were maintained in RPMI medium (Wako, Osaka, Japan) with 10% fetal bovine serum (HyClone; GE Healthcare Life Sciences, Logan, UT, USA) and 1% penicillin–streptomycin (Gibco; Thermo Fisher Scientific, Waltham, MA, USA) at 37°C in a humidified chamber with 5% CO_2.

2.2 *Irradiation*

Irradiation was performed using an X-ray irradiation machine (Titan-225S; Shimadzu Corporation, Kyoto, Japan) at a rate of 1.3 Gy/min. The doses of irradiation were 2 Gy, 4 Gy, 6 Gy, and 8 Gy based on the common fractionation dose for radiotherapy.

2.3 Clonogenic assay

Clonogenic assay was performed to assess the survival of both cells. Cells were cultured in a 10 cm culture dish at a density of 1×10^6 cells/10 ml medium. After the cells had reached 80% confluence, CHEK-1 cells and TE-8 cells were irradiated with 2 Gy, 4 Gy, 6 Gy and 8 Gy doses of X-ray irradiation. Immediately following irradiation, the cells (500 cells/4 ml medium) were seeded in 25 cm² Falcon tissue culture flasks. Following 14 days of culture at 37°C and 5% CO_2, the cells were washed with PBS (Phosphate Buffer Saline), then fixed with 99.5% ethanol, and stained with 0.5% crystal violet in H_2O: methanol (1:1) for 30 min at room temperature. The cells were then washed with tap water and air-dried. The total number of colonies containing >50 cells was counted using a binocular light microscope (Olympus Corporation, Tokyo, Japan). The Plating Efficiency (PE) and Survival Fraction (SF) were calculated using the following equations (Buch et al., 2012, Franken et al., 2006):

$$PE = \frac{\text{Number of colonies formed}}{\text{Number of cells seeded}} \times 100\%$$

$$SF = \frac{\text{Number of colonies formed post irradiation}}{\text{Number of cells seeded} \times PE} \times 100\%$$

2.4 Statistical analysis

Data are presented as the mean ± standard error of the mean from three independent experiments. The differences between multiple variables were analyzed by one-way analysis of variance (ANOVA) with Dunnett's pairwise comparison for the post-hoc analysis. The difference between two means was analyzed by the t-test. A probability of $p < 0.05$ was considered significant. All statistical analyses were performed using the EZR statistical software, an open-source statistical software program that is based on R and R commander (Kanda, 2013).

3 RESULTS

3.1 Survival of CHEK-1 cells

Figure 1 presents the SF of CHEK-1 cells treated with X-ray irradiation of doses 0 Gy, 2 Gy, 4 Gy, 6 Gy, and 8 Gy.

The percentages of SF were 100%, 67%, 34%, 12%, and 1%, respectively. The SF between the control (0 Gy) and irradiated groups is statistically significantly different ($p < 0.05$).

3.2 Survival of TE-8 cells

Figure 2 shows the SF of TE-8 cells treated with X-ray irradiation of doses 0, 2 Gy, 4 Gy, 6 Gy, and 8 Gy. The percentages of SF were 100%, 86%, 36%, 7%, and 3%, respectively. The SF between the

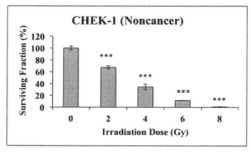

n=3 *** P<0.001

Figure 1. Surviving fraction of CHEK-1 cells.

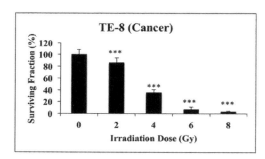

n=3 *** P<0.001

Figure 2. Surviving fraction of TE-8 cells.

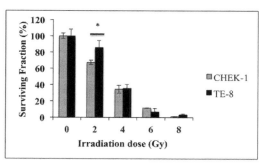

n=3 *P<0.05

Figure 3. Surviving fraction between CHEK-1 and TE-8 cells.

control (0 Gy) and irradiated groups is statistically significantly different (p < 0.05).

3.3 *SF between CHEK-1 and TE-8 cells*

Figure 3 shows the SF between CHEK-1 and TE-8 cells. SF is not significantly different between the two cells (p > 0.05) except for 2Gy X-ray irradiation (p < 0.05).

4 DISCUSSION

This was a preliminary study to observe the effect of different doses of irradiation on noncancerous and cancerous cells. The results indicated that X-ray irradiation starting with a dose of 2 Gy can decrease the survival rate of both noncancerous and cancerous cells, which is statistically significant (P < 0.001). These results are in accordance with the common fractionation dose for cancer treatment with radiation that used 2 Gy as a fractionation dose for the therapy (Fowler, 2010).

Esophageal cells are one of the cells that have fairly high radiosensitivity (Rubin and Casarett, 1968, Murat Beyzadeoglu, 2010); hence, we used these cells for investigating the survival of cells with different doses of irradiation.

The percentage of surviving fraction with a dose of 2 Gy X-ray irradiation for noncancerous cells was lower than that for cancerous cells (67% vs 87%) (P < 0.05). It indicates that cancerous cells are stronger than noncancerous cells against irradiation.

X-ray irradiation starting with a dose of 2Gy can decrease the survival rate of both noncancerous and cancerous cells. Therefore, 2 Gy can be used as an irradiation dose for further study such as the study of radioprotective agents that protect noncancerous cells during radiation treatment or the radiosensitizer study that is intended to enhance the killing of cancer cells while having much less effect on normal tissues.

REFERENCES

Abdulah, R., Faried, A., Kobayashi, K., Yamazaki, C., Suradji, E.W., Ito, K., Suzuki, K., Murakami, M., Kuwano, H. & Koyama, H. 2009. Selenium enrichment of broccoli sprout extract increases chemosensitivity and apoptosis of LNCaP prostate cancer cells. *BMC Cancer*, 9, 414.

Anupama Munshi, M.H., and Raymond E. Meyn 2005. Clonogenic Cell Survival Assay. *In:* Blumenthal, R.D. (ed.) *Chemosensitivity.* Humana Press Inc.

Buch, K., Peters, T., Nawroth, T., Sanger, M., Schmidberger, H. & Langguth, P. 2012. Determination of cell survival after irradiation via clonogenic assay versus multiple MTT Assay--a comparative study. *Radiat Oncol*, 7, 1.

Fowler, J.F. 2010. 21 years of biologically effective dose. *Br J Radiol*, 83, 554–68.

Franken, N.A., Rodermond, H.M., Stap, J., Haveman, J. & Van Bree, C. 2006. Clonogenic assay of cells in vitro. *Nat Protoc*, 1, 2315–9.

Kanda, Y. 2013. Investigation of the freely available easy-to-use software 'EZR' for medical statistics. *Bone Marrow Transplant*, 48, 452–8.

Murat Beyzadeoglu, G.O., Cüneyt Ebruli 2010. *Basic Radiation Oncology*, Springer.

National-Cancer-Institute. *Radiation Therapy for Cancer* [Online]. Available: https://www.cancer.gov/about-cancer/treatment/types/radiation-therapy/radiation-fact-sheet [Accessed 17 October 2016].

Puspitasari, I.M., Abdulah, R., Yamazaki, C., Kameo, S., Nakano, T. & Koyama, H. 2014. Updates on clinical studies of selenium supplementation in radiotherapy. *Radiat Oncol*, 9, 125.

Rubin, P. & Casarett, G.W. 1968. Clinical radiation pathology as applied to curative radiotherapy. *Cancer*, 22, 767–78.

The effect of mesenchymal stem cells on the endothelial cells of diabetic mice

A. Putra
Stem Cell and Cancer Research, Sultan Agung Islamic University, Semarang, Central Java, Indonesia

A. Rahmalita, Y. Tarra, D.H. Prihananti & S.H. Hutama
Undergraduate Student of Medical Faculty, Sultan Agung Islamic University, Semarang, Central Java, Indonesia

N.A.C. Sa'diyah
Department of Internal Medicine, Sultan Agung Islamic University, Semarang, Central Java, Indonesia

ABSTRACT: The damage in pancreatic β cells can be repaired by Mesenchymal Stem Cells (MSCs) because of their differentiation and paracrine potential. The purpose of this study is to find the effect of mouse umbilical cord MSCs ((mUC)-MSCs) on the endothelial cells of pancreatic tissues. The study was conducted on a total of 18 male mice (*Mus musculus*) that were injected with streptozotocin to induce diabetes, and then they were randomly divided into three groups. The treatment group was intraperitoneally injected with MSCs at doses of 1.5×10^5 and 3.0×10^5, named as T1 and T2 (treatment groups 1 and 2), while the control group was injected with PBS. The analysis was carried out after 44 days. The results of the analysis revealed that the number of endothelial cells of pancreatic tissues in the treatment groups was 4.40 ± 1.43 and 6.10 ± 2.33 (T1 and T2 groups), respectively, while that in the control group was 1.27 ± 0.84. The difference was significant ($p < 0.05$) between the treatment groups and the control group, but no significant difference was found between the T1 and T2 groups. It can be concluded that (mUC)-MSCs has an effect on the endothelial cells of pancreatic tissues in diabetic mice.

1 INTRODUCTION

Diabetes mellitus is a condition where insulin production is decreased or a decrease in tissue sensitivity to insulin might cause hyperglycemia and lead to serious complications, especially in blood vessels and nerves, as well as causing damage to pancreatic β cells via Toll-Like Receptors (TLRs 2 and 4). Until now, to keep blood glucose levels under control and to reduce the complication of diabetes, only insulin or an anti-diabetic agent has been used, but they are only a temporary remedy and reduce the symptom (Bluestone et al. 2010). Treatment is needed to repair the damage of pancreatic tissues by activating the surrounding cells or the cells of pancreatic tissues itself into pancreatic β cells.

Islet cell transplantation has been explored as an effective treatment for type 1 diabetes by producing new β cells for controlling blood glucose (Ashcroft & Rorsman. 2012). However, this technique has disadvantages such as lack of sufficient donors and the instant blood-mediated inflammatory reaction (IBMIR) which can lead to the damage of pancreatic β cells of islets (Bennet et al. 2000). Thus, new alternative sources for transplantable islet β cells should be explored. Many studies have focused on mesenchymal stem cells, which are the subpopulation of multipotent cells that can differentiate into other cell types, including pancreatic β endothelial cells (Jones et al. 2007).

MSCs are characterized by their fibroblast-like appearance, and have the ability to differentiate into a specific cell, have colony-forming unit capacity, and can adhere to plastic surfaces (Dominici et al. 2006). MSCs express surface molecules, including CD90, CD73, CD105, CD29, CD44, and CD166, but do not express endothelial or hematopoietic markers, including CD31, CD45, CD43, CD14, CD11b, major histocompatibility complex (MHC) class II molecule, and costimulatory proteins such as CD80, CD86, and CD40 (Dominici et al. 2006; Lv et al. 2014).

MSCs have been used in both experimental models and in the clinical setting as an immunosuppressive treatment and as catalyzers of endothelial cell sprout formation (Johansson et al. 2005). However, the effect of MSCs, particularly derived from the mouse umbilical cord, on the endothelial cells of pancreatic tissues still remains unclear. The purpose of this study is to find the effect of umbilical cord MSCs on the amount of vascular endothelial cells of pancreatic tissues.

2 MATERIALS AND METHODS

2.1 Isolation of (mUC)-MSCs

Mouse umbilical cord (mUC)-MSCs was obtained from the umbilical cord of pregnant female mice (19 days old). Briefly, after the collection of all the UC samples, the samples were rinsed by repeated immersion in Phosphate-Buffered Saline (PBS) (GIBCO, Invitrogen). UC vessels were cleared off and minced finely (0.2–0.5 cm^3) by sterile scissors in a tissue culture dish, and then placed in a 60 mm tissue culture dish with an array of dots spread evenly on the surface of the tissue culture dish containing DMEM low glucose (Sigma-Aldrich, St Louis, MO) supplemented with 10% FBS and 100 IU/ml penicillin/streptomycin (GIBCO, Invitrogen), with the culture medium being changed every 3 days, and then the medium was incubated at 37°C with 5% CO_2. The efficiency of the medium was evaluated by the time required for adherent cells to appear and then reach 80% confluence. All procedures were approved by the Ethics Committee for Animal Research of the Unissula Medical School.

2.2 Induction of experimental diabetes

A total of 18 male mice (*Mus musculus*) aged 10 weeks were intraperitoneally injected with 40 mg/kg streptozotocin (STZ; Sigma-Aldrich) for 5 consecutive days to induce diabetes. STZ was diluted in sodium citrate buffer (pH 4.5). Blood samples were taken from the tail vein of non-fasting mice, and glucose levels were determined with a glucometer system Accu-Chek Active (Roche Diagnostics, Abbott Park, IL, USA). Mice were considered diabetic when blood glucose levels exceeded 250 mg/dl in two consecutive determinations. All animal procedures were approved by the Ethics Committee for Animal Research of the Unissula Medical School.

2.3 In vitro osteogenic differentiation

To differentiate into the osteogenic cell, at passage 3/5, putative cells were seeded in a six-well culture plate at a density of $5 \times 10^3/1 \times 10^4$ cells/well in 24-well plates and cultured in DMEM for 24 h to allow cell adhesion. The adhered cells were treated with an osteogenic induction medium supplemented by 10 mmol/L of β-glycerophosphate, 0.1 μM dexamethasone, 50 μmol/L of ascorbate-2-phosphate (all purchased from Sigma-Aldrich, St Louis, MO), and 10% (v/v) FBS in DMEM. Osteogenic differentiation was confirmed by Alizarin Red staining after 21 days of induction for detecting calcium deposition. In brief, cells were rinsed with PBS and fixed to incubate in ice-cold 70% ethanol (v/v) for 1 h at Room Temperature (RT), and rinsed with distilled water three times. Then, 1 ml of 2% (w/v) Alizarin Red S (pH 4.1–4.3) solution was added and incubated at RT for 30 min, and subsequently removed and rinsed with distilled water four times.

2.4 Immunophenotyping of (mUC)-MSCs

Characterization of (mUC)-MSCs was carried out by immunocytochemistry using MSC-positive markers. Briefly, expanded (mUC)-MSCs were passaged after reaching 60 to 80% confluency, and grown on glass coverslips, and then fixed with 4% paraformaldehyde in 90% ethanol for 15 minutes at 4°C. Subsequently, cells were incubated with primary antibody CD 73 and CD 105 (1:100; BD Pharmingen, San Diego, CA USA) for 60 min at room temperature, washed with PBS for 10 min, labeled with secondary antibody (1:2500 dilution) for 15 min at room temperature, counterstained with DAB (Santa Cruz Biotech), and then visualized by a phase-contrast microscope.

2.5 (mUC)-MSC injection

Diabetic mice (n = 6/group) in the treatment groups were injected intraperitoneally (intrasplenic) with MSCs at doses of 1.5×10^5 and 3.0×10^5 for 20 days, and the control group received an intrasplenic injection of PBS (n = 6). For peritoneal injections of MSCs or PBS, mice were anesthetized with a mixture of ketamine (Ketamine, Brazil) and xylazine (Dopaser, Brazil) and then received a single dose of 70 μl MSC-containing solution or PBS microinjection. Bleeding was controlled using cotton swabs and by the local application of fibrin sealant. Intraperitoneal injections of tramadol hydrochloride (30 mg/kg, Brazil) were used for pain relief every 12 hours for 3 consecutive days. Non-fasting glucose was monitored every 5 days by using the glucometer system Accu-Chek Active (Roche Diagnostics). Mice were killed 35 days after receiving MSC or PBS injection, and the pancreatic samples were collected and histological samples were prepared to determine the number of vascular endothelial cells in pancreatic tissues.

2.6 Histological analysis

For histological analysis, pancreatic tissue was removed, fixed in 10% neutral buffered formalin, embedded in paraffin, the sections (5 μm) were stained with hematoxylin and eosin (H&E), and analyzed under a light microscope.

2.7 Statistical analysis

Data are presented as mean ± Standard Deviation (SD). Statistical comparisons were made by one-way analysis of variance with Tukey's *post hoc* test. $P < 0.05$ was considered as significant.

3 RESULTS

To characterize MSCs, we evaluated cell morphology, the expression of surface markers, and the *in vitro* differentiation potential. MSCs appeared as typical monolayers of spindle-shaped fibroblast-like cells (Figure 1a), with the ability to adhere to plastic surfaces during *in vitro* expansion. MSC samples presented typical MSC phenotypes of CD73 (Figure 1a) and CD105 (Figure 1b). To evaluate the *in vitro* osteogenic differentiation potential, MSCs were cultured for 21 days with a specific medium to induce differentiation into adipocytes. Cytoplasmic lipid vesicles were detected by Alizarin Red Staining (Figure 2).

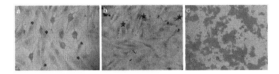

Figure 1. MSCs appeared as spindle-shaped fibroblast-like cells (a). MSCs presented typical MSC phenotypes of CD73 (a) and CD105 (b). *In vitro* osteogenic differentiation, marked in red color, was detected by Alizarin Red staining (c).

Figure 2. Endothelial pancreatic islet cells of the control group (a) and the treatment groups at a dose of 1.5×10^5 (b) and at a dose of 3.0×10^5 (2c). The difference between the control and treatment groups is shown by the increase in endothelial pancreatic islet cell proliferation.

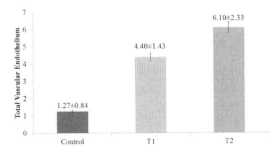

Figure 3. Graph showing endothelial cell proliferation in pancreatic cells. The treatment groups differed significantly from the control group ($p < 0.05$), but there was no significant difference found between the T1 (MSCs at a dose of 1.5×10^5) and T2 (MSCs at a dose of 3.0×10^5) groups.

After the 7th day, mice were killed for histopathological examination. We found that the number of vascular endothelial cells in the pancreatic tissues of the treatment groups was 4.40 ± 1.43 (T1) and 6.10 ± 2.33 (T2) compared with the control group, which was only 1.27 ± 0.84 (Figure 3). These results indicated that all the mean values of the treatment groups were significantly different from that of the control group, showing that the mesenchymal stem cell has the ability to improve vascular endothelial cells in the pancreatic tissues of diabetic mice.

4 DISCUSSION

MSCs have been considered as a promising therapeutic approach for inflammatory diseases and Autoimmune Diseases (AIDs) due to their immunomodulatory properties. The ability of MSCs to suppress the immune response suggests a possible role of these cells in promoting tolerance to AID, and supports their application in the treatment of T1D (type 1 diabetes mellitus).

In our study, MSCs appeared as fibroblast-like cells, expressed surface markers, i.e. CD73 and CD105, and had the osteogenic differentiation potential, which are consistent with the results of previous studies (Dominici et al. 2006; Lv et al. 2014). The results of this study showed that administration of MSCs was able to increase the amount of vascular endothelial tissues in the pancreas in all the treatment groups compared with the control group. However, there was no difference found between the T1 and T2 groups, showing that the doses of MSCs did not correlate with the proliferation of endothelial cells in pancreatic tissues.

Theoretically, MSCs has a tendency to migrate to areas of injury and has the multi-differentiation potential, particularly differentiating into vascular endothelial cells, which is in line with our study. In addition to differentiation, MSCs are thought to release some growth factors, particularly Vascular Endothelial Growth Factor (VEGF) (Gnecchi et al. 2008) as a mediator of the secretion signal of pancreatic β cells to endothelial cells and as a main molecule for inducting the formation of capillary tissues in the islets of Langerhans (Konstantinoval & Lammert. 2004).

The role of MSCs is as follows: they would first enter the bloodstream in advance and produce hematopoietic cells, and bind with E-selectin for rolling and adhesion, and then they migrate and penetrate into endothelial cells and toward the damaged area of pancreatic β cells, thereby differentiating, replicating, and trans-differentiating through a paracrine mechanism in the injured area (Ezquer et al. 2012; Korbling & Estrov. 2003; Wan et al. 2013).

5 CONCLUSION

This study indicated that (mUC)-MSCs have the potential to increase the amount of vascular endothelial cells in pancreatic tissues.

ACKNOWLEDGMENTS

The authors gratefully acknowledge the Stem Cell and Cancer Research Laboratory of Medical Faculty of Unissula for their support and facility.

REFERENCES

Ashcroft, F.M. & Rorsman, P. 2012. Diabetes mellitus and the β cell: the last ten years. *Cell* 148: 1160–1171.

Bennet, W., Sundberg, B., Lundgren, T., Tibell, A., Groth, C.G., Richards, A., White, D.J., Elgue, G., Larsson, R., Nilsson, B., & Korsgren, O. 2000. Damage to porcine islets of Langerhans after exposure to human blood in vitro, or after intraportal transplantation to cynomologus monkeys: protective effects of sCR1 and heparin. *Transplantation* 69 (5): 711–9.

Bluestone, J.A., Herold, K., & Eisenbarth, G. 2010. Genetics, pathogenesis and clinical interventions in type 1 diabetes. *Nature* 464: 1293–1300.

Dominici, M., Le Blanc, K., Mueller, I., Slaper-Cortenbach, I., Marini, F., Krause, D., Deans, R., Keating, A., Prockop, Dj., & Horwitz, E. 2006. Minimal criteria for defining multipotentmesenchymal stromal cells. The International Society for Cellular Therapy position statement. *Cytotherapy* 8 (4): 315–7.

Ezquer, F., Ezquer, M., Contador, D., Ricca, M., Simon, V., & Conget, P. 2012. The antidiabetic effect of mesenchymal stem cells is unrelated to their transdifferentiation potential but to their capability to restore Th1/Th2 balance and to modify the pancreatic microenvironment. *Stem Cells* 30 (8): 1664–74.

Gnecchi, M., Zhang, Z., Ni, A., & Dzau, V.J. 2008. Paracrine mechanisms in adult stem cell signaling and therapy. *Circ Res* 103 (11): 1204–219.

Johansson, U., Elgue, G., Nilsson, B., & Korsgren, O. 2005. Composite islet-endothelial cell grafts: a novel approach to counteract innate immunity in islet transplantation. *Am J Transplant* 5 (11): 2632–9.

Jones, S., Horwood, N., Cope, A., & Dazzi, F. 2007. The antiproliferative effect of mesenchymal stem cells is a fundamental property shared by all stromal cells. *J Immunol* 179 (5):2824–31.

Konstantinoval. & Lammert, E. 2004. Microvascular development: learning from pancreatic islets. *Bioessays* 26 (10): 1069–75.

Korbling, M. & Estrov, Z. 2003. Adult stem cells for tissue repair-a new theropeutic concept?. *N Eng J Med* 349: 570–582.

Lv, F.J., Tuan, R.S., Cheung, K.M., & Leung, V.Y. 2014. Concise review: the surface markers and identity of human mesenchymal stem cells. *Stem Cells* 32 (6): 1408–19.

Wan, J., Xia, L., Liang, W., Liu, Y., & Cai, Q. 2013. Transplantation of bone marrow-derived mesenchymal stem cells promotes delayed wound healing in diabetic rats. *J Diabetes Res* 2013 (10): 647107.

Advances in Biomolecular Medicine – Hofstra, Koibuchi & Fucharoen (Eds)
© 2017 Taylor & Francis Group, London, ISBN 978-1-138-63177-9

Effects of selenium on SePP and Apo B-100 Gene expressions in human primary hepatocytes

M. Putri
Department of Biochemistry, Faculty of Medicine, Universitas Islam Bandung, Bandung, Indonesia
Department of Public Health, Gunma University Graduate School of Medicine, Gunma, Japan
Department of Medicine and Biological Science, Gunma University Graduate School of Medicine, Gunma, Japan

N. Sutadipura & S. Achmad
Department of Biochemistry, Faculty of Medicine, Universitas Islam Bandung, Bandung, Indonesia

C. Yamazaki, S. Kameo & H. Koyama
Department of Public Health, Gunma University Graduate School of Medicine, Gunma, Japan

M.R.A.A. Syamsunarno
Department of Medicine and Biological Science, Gunma University Graduate School of Medicine, Gunma, Japan
Department of Biochemistry, Faculty of Medicine, Universitas Padjadjaran, Jatinangor, Indonesia

T. Iso & M. Kurabayashi
Department of Medicine and Biological Science, Gunma University Graduate School of Medicine, Gunma, Japan

ABSTRACT: The effects of selenium on Selenoprotein P (SePP) and apolipoprotein B-100 (apoB-100) expressions were observed in Human Primary Hepatocytes (Hc cells) under basal state condition, to represent the human liver in healthy conditions. The Hc cells were cultured in a medium supplemented with 0–200 nM sodium selenite. The effects of sodium selenite supplementation on SePP and apoB-100 genes were measured by real-time PCR, respectively. mRNA expressions of SePP and apoB were up-regulated after treatment with sodium selenite, and reached the optimum effects at the same dose (50 nM). These results suggest that sodium selenite supplementation might play a role in SePP and apoB-100 production in Hc cells under basal state condition.

1 INTRODUCTION

Selenium is an essential micronutrient for human health, which exists in various chemical forms, and the physiological effects of the different chemical forms vary considerably. Sodium selenite, an inorganic selenium form, is used as a supplement; recently, it is being increasingly consumed by healthy people to delay the onset or progression of age-related degenerative diseases, including cardiovascular disease (Actis-Goretta et al., 2004). Selenium functions for human were through selenoproteins (Brenneisen et al., 2005). Selenoprotein p (SePP) is the main selenium transport form in plasma (Burk and Hill, 2009), tissues collect "long isoform" SePP from plasma by apoER2-mediated endocytosis and use its selenium for the synthesis of selenoproteins (Kurokawa et al., 2012). The main source of plasma SePP is in the liver (Schweizer et al., 2005), even though lower amounts SePP mRNA also present in many other tissues, suggesting that SePP is widely expressed (Hoffmann et al., 2007). Base on this fact, production of SePP by hepatocytes is vital to selenium homeostasis in the organism (Hill et al., 2012).

Cholesterol in the blood is mainly carried by Low-Density Lipoprotein (LDL) and most significantly associated with atherosclerotic plaque formation (Gouni-Berthold and Sachinidis, 2004). Apolipoprotein B (apo B) is correlated with LDL particles, a study by Dhingra and Bansal, 2005 demonstrated that in mice, selenium deficiency caused increased of apo B expression during experimental hypercholesterolemia and Selenium supplementation leads to down-regulation of apo B expression (Dhingra and Bansal, 2005). Apolipoprotein B 100 (apo B-100) is apo B that is consists of the ligand-binding domain for binding of LDL to LDL-R site (Segrest et al., 2001). Several studies proposed that apoB-100 levels might be a better indicator than total or LDL cholesterol levels for measuring the concentration of atherogenic lipoprotein particles (Davidson, 2012) because most of the studies demonstrated that one molecule of apo B-100 exists per lipoprotein particle, hence the quantity of apo B-100 in fasting plasma predicts the sum of LDL and VLDL particles (Vega and Grundy, 1990, Harper and Jacobson, 2010). Based on this facts, there was lacking data for the efficacy of selenium supplementation for suppressing Apo B-100 in the healthy human.

Hence in view of all the above-stated findings, the present study is aimed to explore the effect of selenium supplementation on SePP and apo B-100 expressions in Human Primary Hepatocytes (Hc cells) under basal state condition. We used non-malignant cells for a better representation of the human liver in healthy conditions, to mimic healthy individuals receiving selenium supplementation in an in vitro system. We hypothesized that sodium selenite supplementation would up-regulate SePP and down-regulate apo B-100 expressions in the normal human liver.

2 MATERIAL AND METHOD

2.1 *Cell culture*

The Hc cells were kindly provided by Prof. Takeaki Nagamine, who purchased them from the Applied Cell Biology Research Institute (Kirkland, WA). The cells were maintained as described in a previous study (Hayakawa and Nagamine, 2014) and incubated on dishes coated with type I collagen. For the experiments, cells were seeded at a concentration of 1.5×10^6 cells per 100-mm dish and incubated for 0–72 h. Hc cells were cultured under the same conditions for the selenium-supplemented groups, except that sodium selenite was added to the medium at the specified concentrations.

In our previous research, the selenium concentration in FBS was measured and determined it to be 257.65 nM. Hence, the selenium concentration of the culture medium with 10% FBS was 25.76 nM. The concentrations of selenium in the experiments are the supplemented concentrations to the selenium contained in the FBS. Therefore, the culture medium of the control group (0 nM) contained 25.76 nM of selenium from the FBS (Putri et al., 2014).

2.2 *Quantitative real-time polymerase chain reaction analysis*

After incubation in various concentrations of sodium selenite (0, 25, 50, 100, and 200 nM) for 72 h, total RNA was isolated from Hc cells using TRIzol reagent (Invitrogen, CA, USA). RNA was prepared by reverse transcription using oligo-dT

Table 1. Primers for quantitative real-time PCR.

Primer	Sequences
hApo B-100	TCGCCTGCCAAACTGCTTC (F)
	CATTGTGCCTGTGTTCCATTC (R)
hGAPDH	ACCACATCCATGCCATCAC (F)
	TCCACCACCCTGTTGCTGTA (R)
hSePP	TTCGGGCAGAGGA-GAACA (F)
	CTGGCACTGGCTTCTGTG (R)

*hApo B-100, human apolipoprotein B-100; hGAPDH, human glyceraldehyde-3-phosphatase dehydrogenase; hSePP, human selenoprotein P; F, forward; R, reverse.

and dNTP, and each sample was processed with the RT-PCR kit (TAKARA, Japan). Quantitative real time-PCR was performed using the SYBR Green PCR Master Mix (Applied Biosystems, CA, USA) according to the manufacturer's instructions, and then evaluated using the Light Cycler 480 Real-Time PCR system (Roche, CA, USA). The expression level of the target gene was normalized against GAPDH mRNA levels. The sequences of primers for quantitative real-time PCR used in this study are listed in Table 1.

2.3 *Statistical analysis*

Statistical analysis was performed using 1-way analysis of variance with Dunnett's posthoc multiple comparison tests to compare between "no treatment" as the control group and each experimental group. A p-value < 0.05 was considered statistically significant. The statistical analysis of the data was performed with IBM SPSS (version 21.0 for Windows, IBM, NY, USA).

3 RESULT AND DISCUSSION

Based on a previous study (Hoefig et al., 2011) we used 72 h as the optimal incubation time for sodium selenite supplementation. Selenium has a very narrow therapeutic dose window (Rayman, 2008); therefore in our previous research we confirmed the selenium dose and associated cell viability using a Methyl Thiazolyl Tetrazolium assay (MTT) assay, the Hc cells showed no inhibition of cell proliferation after 72-h incubation with doses ≤ 200 nM (Putri et al., 2014).

In this study, we demonstrated that in Hc cells, sodium selenite supplementation significantly increased mRNA expressions of SePP and apo B-100 (Figure 1), suggesting that selenium induced SePP and apo B-100 expression from the transcription level. Interestingly, both reached the peak at the same dose of sodium selenite (50 nM)

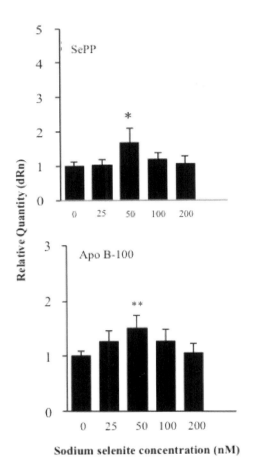

Figure 1. mRNA Expressions of sePP and apo B-100 in Hc cells after treatment with sodium selenite as analyzed by real-time PCR.

Our result of increasing mRNA expression of SePP after sodium selenite supplementation in Hc cells was in agreement with the previous study. Hill, et.al, 2012 demonstrated that selective deletion of SePP in hepatocytes decreased plasma SePP concentration in mice fed a selenium-adequate diet leads to major changes in whole-body selenium metabolism. Those changes in metabolism disrupt selenium supply to extra-hepatic tissues were deteriorated in dietary selenium deficiency. (7). The liver has a greater supply of the element than do other tissues, this makes it is well appropriate to regulate whole-body selenium metabolism (Kato et al., 1992). Sodium selenite, an inorganic form of selenium, is well established as a dietary supplement, and its consumption has increased over the past few years (15).

Next, our study showed sodium selenite supplementation significantly increased mRNA expressions of apo B-100 in Hc cells, and reached the saturation at 50 nM sodium selenite. There was a previous study that during experimental hypercholesterolemia, selenium deficiency caused increased of apo B expression and Selenium supplementation leads to down-regulation of apo B expression (9). Combination with our present result, it suggesting that in the healthy human, supplementation of sodium selenite will induce apo B-100 concentration and to suppress this gene may need a stronger transcriptional suppressor, such as hypercholesterolemia. The mechanism why apo B-100 was induced by selenium in normal human liver cells is still unexplained. Further in vivo studies are necessary to explore potential roles of selenium especially in healthy condition.

4 CONCLUSION

In Hc cells that represent normal human liver cells, selenium supplementation up-regulated SePP and apo B-100 expressions.

ACKNOWLEDGMENTS

This study was supported by the GP 2012 grant (Gunma University, Japan). The authors would like to express their gratitude to Prof. Takeaki Nagamine for his kind gift of the Hc cells, and to Chiho Yoshizawa and Miki Matsui for technical assistance.

All the authors have read the manuscript and have agreed to submit it in its current form for consideration for publication in the journal. There are no conflicts of interest to declare.

REFERENCES

Actis-Goretta, L., Carrasquedo, F. & Fraga, C. G. (2004) The regular supplementation with an antioxidant mixture decreases oxidative stress in healthy humans. Gender effect. *Clin Chim Acta*, 349, 97–103.

Brenneisen, P., Steinbrenner, H. & Sies, H. (2005) Selenium, oxidative stress, and health aspects. *Mol Aspects Med*, 26, 256–267.

Burk, R. F. & Hill, K. E. (2009) Selenoprotein P-expression, functions, and roles in mammals. *Biochim Biophys Acta*, 1790, 1441–7.

Davidson, M. H. (2012) Low-density lipoprotein cholesterol, non-high-density lipoprotein, apolipoprotein, or low-density lipoprotein particle: what should clinicians measure? *J Am Coll Cardiol*, 60, 2616–7.

Dhingra, S. & Bansal, M. P. (2005) Hypercholesterolemia and apolipoprotein B expression: regulation by selenium status. *Lipids Health Dis*, 4, 28.

Gouni-Berthold, I. & Sachinidis, A. (2004) Possible non-classic intracellular and molecular mechanisms of

LDL cholesterol action contributing to the development and progression of atherosclerosis. *Curr Vasc Pharmacol,* 2, 363–70.

Harper, C. R. & Jacobson, T. A. (2010) Using apolipoprotein B to manage dyslipidemic patients: time for a change? *Mayo Clin Proc,* 85, 440–5.

Hill, K. E., Wu, S., Motley, A. K., Stevenson, T. D., Winfrey, V. P., Capecchi, M. R., Atkins, J. F. & Burk, R. F. (2012) Production of selenoprotein P (Sepp1) by hepatocytes is central to selenium homeostasis. *J Biol Chem,* 287, 40414–24.

Hoefig, C. S., Renko, K., Köhrle, J., Birringer, M. & Schomburg, L. (2011) Comparison of different selenocompounds with respect to nutritional value vs. toxicity using liver cells in culture. *J Nutr Biochem,* 22, 945–955.

Hoffmann, P. R., Hoge, S. C., Li, P. A., Hoffmann, F. W., Hashimoto, A. C. & Berry, M. J. (2007) The selenoproteome exhibits widely varying, tissue-specific dependence on selenoprotein P for selenium supply. *Nucleic Acids Res,* 35, 3963–73.

Kato, T., Read, R., Rozga, J. & Burk, R. F. (1992) Evidence for intestinal release of absorbed selenium in a form with high hepatic extraction. *Am J Physiol,* 262, G854–8.

Kurokawa, S., Hill, K. E., Mcdonald, W. H. & Burk, R. F. (2012) Long isoform mouse selenoprotein P (Sepp1) supplies rat myoblast L8 cells with selenium via endocytosis mediated by heparin binding properties and apolipoprotein E receptor-2 (ApoER2). *J Biol Chem,* 287, 28717–26.

Putri, M., Yamazaki, C., Syamsunarno, M., Puspitasari, I., Abdulah, R., Kameo, S., Iso, T., Kurabayashi, M. & Koyama, H. (2014) Effects Of Sodium Selenite Supplementation On Preβ -high-density Lipoprotein Formation-related Proteins in Human Primary Hepatocytes. *International Journal of Food and Nutritional Sciences,* 3, 16–22.

Rayman, M. P. (2008) Food-chain selenium and human health: emphasis on intake. *Br J Nutr,* 100, 254–268.

Schweizer, U., Streckfuss, F., Pelt, P., Carlson, B. A., Hatfield, D. L., Kohrle, J. & Schomburg, L. (2005) Hepatically derived selenoprotein P is a key factor for kidney but not for brain selenium supply. *Biochem J,* 386, 221–6.

Segrest, J. P., Jones, M. K., De Loof, H. & Dashti, N. (2001) Structure of apolipoprotein B-100 in low density lipoproteins. *J Lipid Res,* 42, 1346–67.

Vega, G. L. & Grundy, S. M. (1990) Does measurement of apolipoprotein B have a place in cholesterol management? *Arteriosclerosis,* 10, 668–71.

Advances in Biomolecular Medicine – Hofstra, Koibuchi & Fucharoen (Eds)
© 2017 Taylor & Francis Group, London, ISBN 978-1-138-63177-9

Inhibition of cAMP synthesis abolishes the impact of curcumin administration in the skeletal muscle of rodents

H.R.D. Ray
Universitas Pendidikan Indonesia, Bandung, West Java, Indonesia

K. Masuda
Kanazawa University, Kanazawa, Ishikawa, Japan

ABSTRACT: Our previous experiments have shown that exercise increases cyclic Adenosine Monophosphate (cAMP) levels followed by activated Protein Kinase A (PKA) and the downstream target of PKA including phosphorylation of LKB-1 and CREB. The purpose of the present study was to examine whether the effect of the PKA inhibitor H89 abolishes the effect of curcumin on the regulation of mitochondrial biogenesis. Male Wistar rats aged 8 weeks were randomized into two groups: exercise group and non-exercise group. They were further divided into the following groups: Control group (DMSO), curcumin group (100 mg/kg-BW/day), H89 group (20 mg/kg-BW/day), and curcumin + H89 group (intraperitoneally injected for 3 days). The exercise regimen was swimming for 2 hours/day for 3 days. Western Blot (WB) and Immunoprecipitation (IP) analyses were carried out to assess protein and enzyme activities. We found that the administration of curcumin alone and in combination with exercise increased the phosphorylation of LKB-1, CREB, and COX-IV expression. However, interestingly, H89 abolished these effects. The present results indicate that inhibition of cAMP synthesis abolishes the impact of curcumin administration in the skeletal muscle of rodents.

1 INTRODUCTION

Skeletal muscle is a highly malleable tissue, capable of considerable metabolic and morphological adaptations in response to repeated bouts of contractile activity (i.e. exercise). It is well established that chronic contractile activity, in the form of repeated bouts of endurance exercise, usually interspersed with recovery periods, results in the altered expression of a wide variety of gene products, leading to an altered muscle phenotype with improved fatigue resistance. This improved endurance is highly correlated with the increase in muscle mitochondrial density and enzyme activity, referred to as "mitochondrial biogenesis" (Nourshahi et al., 2012). The ubiquitous second messenger cyclic AMP (cAMP) and its cellular effector Protein Kinase A (PKA) constitute one of the most widely studied signaling cascades. In both mammalian cells and yeast, the regulation of mitochondrial biogenesis clearly involves the cAMP signaling pathway (Yoboue et al., 2012). Protein Kinase A (PKA), also known as the cAMP-dependent enzyme, is a well-studied, downstream effector of cAMP and is activated only in the presence of cAMP. Activated PKA phosphorylates a number of other proteins including cAMP Response Element Binding protein (CREB) and Liver Kinase B1 (LKB-1), which induces PGC-1α to regulate mitochondria biogenesis (Than et al., 2011, Veeranki et al., 2011).

Several polyphenols have been shown to activate cAMP, which are being deeply investigated as potential inducers of mitochondrial biogenesis (Park et al., 2012, Chowanadisai et al., 2010). Curcumin is a polyphenolic compound with medicinal properties found in *Curcuma longa* L., turmeric, a popular culinary spice used in both vegetarian and non-vegetarian foods. Our previous results have shown that curcumin have the ability to increase mitochondrial biogenesis. Furthermore, when combined with endurance training, curcumin increases mitochondrial biogenesis in skeletal muscle. This study suggested that cAMP may play an important role in the effect of curcumin on the induction of mitochondrial biogenesis in skeletal muscle; however, it remains unknown whether the molecular mechanism of cAMP is involved in the regulation of mitochondrial biogenesis. H89 is identified as a selective and potent inhibitor of Protein Kinase A (PKA). PKA is a ubiquitous cellular kinase that phosphorylates serine and threonine residues in response to cAMP. It is well established the PKA signaling cascade is more important for cellular function and accounts for the profound interest in a specific high-affinity inhibitor.

In the present study, we investigated whether cAMP plays an important role in the effect of curcumin, alone or in combination with exercise, on the regulation of mitochondrial biogenesis in gastrocnemius skeletal muscle. We predicted that the effect of curcumin administered alone and in combination with exercise on the induction of mitochondrial biogenesis through the synthesis of cAMP will be abolished by the PKA inhibitor H89.

2 MATERIALS AND METHODS

2.1 Animal experiments

A total of 36 male Wistar rats (body weight 282–390 g) aged 10 weeks were used in this study (six rats per group). The animals were housed in an air-conditioned room and exposed to a 12-h light–dark photoperiod. They had *ad libitum* access to a standard diet (Oriental Yeast, Tokyo, Japan) and water. The animals were randomly divided into two groups: exercise group and non–exercise group. Subsequently, the two groups were further divided as follows: control group (only DMSO), curcumin group, H89 group (PKA inhibitor), and curcumin + H89 group. All animals were injected intraperitoneally (i.p.) once per day for 3 days with curcumin (100 mg/kg-BW/day) dissolved in dimethyl sulfoxide (DMSO) or in the same volume of DMSO (vehicle alone) and H89 (PKA inhibitor, 20 mg/kg-BW/day). The exercise training group swam 2 h/day in four 30-min bouts separated by 5 min of rest. After the first 30-min bout, a weight equal to 2% of body weight was tied to the body of the rat. The rats swam with the attached weight in the remaining three exercise bouts. All rats swam in a barrel filled to a depth of 50 cm with a swimming area of 190 cm^2/rat (Kawanaka et al., 1997). The rats performed the above swimming protocol once per day for 3 days.

2.2 Nuclear fraction preparation

Animals were anesthetized with 50 mg pentobarbital sodium per 100 g of body weight at 1 hour after the last endurance exercise session. For biochemical studies, gastrocnemius muscle was isolated. Nuclear proteins were isolated using a modified version of the protocol established by Blough (Blough et al., 1999) and divided into 1-ml PBS aliquots.

2.3 Immunoprecipitation and western blotting

Lysis buffer (20 mM Tris-HCl [pH 7.4], 50 mM NaCl, 250 mM sucrose, 50 mM NaF, 5 mM sodium pyrophosphate, 1 mM DTT, 4 mg/l leupeptin, 50 mg/l trypsin inhibitor, 0.1 mM benzamidine, and 0.5 mM PMSF) was used for immunoprecipitation.

Dynabeads (Invitrogen, CA, USA) were added to the samples of nuclear proteins, and incubated at 4°C for 1 h. Dynabeads were then collected using magnets, and normal mouse immunoglobulin G (nIgG; Santa Cruz Biotechnology, CA, USA) was added to the supernatants and incubated at 4°C overnight. The samples were centrifuged for 30 s × 7000 g at 4°C, and aliquots of this supernatant were incubated with primary antibodies against PGC-1α and mouse IgG at 4°C overnight. Dynabeads were then added and incubated at 4°C for 2 h. The pellets were collected using magnets and washed with PBS. The final pellet from immunoprecipitation (IP) was visualized and the expression of proteins was compared by western blotting. Western blot analysis was performed as described previously (Furuichi et al., 2010). Briefly, equal amounts of protein samples were loaded onto SDS–PAGE gel and transferred onto a polyvinylidene fluoride (PVDF) membrane. The membrane was then incubated in blocking buffer, and subsequently with COX-IV (1:1000 dilution; Abcam, Cambridge, England), Phospho LKB-1 (S 428) (1:500 dilution; Abcam, Cambridge, England), LKB-1 (1:500 dilution; Abcam, Cambridge, England), Phospho CREB (Ser 133) (1:500 dilution; Abcam, Cambridge, England), CREB + CREM (1:500 dilution; Abcam, Cambridge, England), anti-β-actin (1:1000 dilution; Abcam, Cambridge, England), GAPDH (1:1000 dilution; Abcam, Cambridge, England), and lamin (1:1000 dilution; Santa Cruz Biotechnology, CA, USA). The signal intensity was quantified using imaging software (Image J, version 1.46; NIH, Maryland, USA).

2.4 Statistical analysis

One-way ANOVA was used to assess the effect of curcumin administered alone and in combined with exercise on the phosphorylation of LKB-1, CREB, and COX-IV expression. The Tukey–Kramer *post hoc* test was used for analysis to identify of difference between the low-dose and high-dose curcumin groups. All data are expressed as mean ± Standard Deviation (SD). The level of significance was established at $P < 0.05$.

3 RESULTS

3.1 PKA inhibitor (H89) diminished the effect of curcumin treatment on the phosphorylation of LKB-1 and CREB

In order to determine whether cAMP plays an important role in the effect of curcumin on the induction of mitochondrial biogenesis, we examined the downstream target proteins of cAMP-dependent protein kinase (PKA) by using the

PKA inhibitor H89. The present result indicated that curcumin without exercise increased the phosphorylation of LKB-1 and CREB (Figs. 1 and 2) in gastrocnemius skeletal muscle, but H89 abolished this effect. Furthermore, our result showed that curcumin treatment increased the effect of exercise that in turn increased the phosphorylation of LKB-1 and CREB (Figs. 2 and 3), but H89 abolished this effect. These results suggested that the administration of curcumin alone and in combination with exercise increased the downstream target of PKA including phosphorylation of LKB-1 and CREB, and when the rats were treated with the PKA inhibitor (H89), the effect of curcumin was abolished.

3.2 PKA inhibitor (H89) inhibited the effect of curcumin on the mitochondrial marker COX-IV

In order to determine the effect of curcumin on the mitochondrial marker, we examine the effect of curcumin treatment on COX-IV expression. Indeed, our result indicated that the administration of curcumin alone and in combination with exercise increased the protein expression of COX-IV, but H89 abolished this effect (Fig. 3).

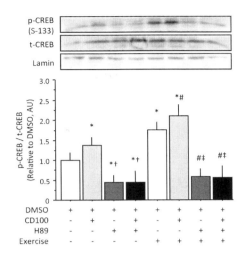

Figure 2. PKA inhibitor H89 abolished the effect of curcumin treatment on the phosphorylation of CREB (S-133) in gastrocnemius skeletal muscle. Values are mean ± SD (n = 6 per group). DMSO = control. CD 100 = curcumin 100 mg/kg-BW/day in DMSO. *: significantly different from the DMSO without exercise group (P < 0.05). #: significantly different from the DMSO + exercise group (P < 0.05). †: significantly different from the curcumin without exercise group (P < 0.05). ‡: significantly different from the curcumin with exercise group.

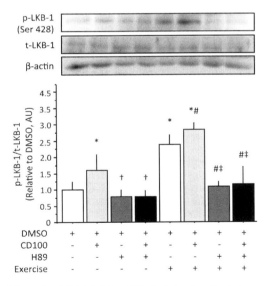

Figure 1. PKA inhibitor H89 abolishes the effect of curcumin treatment on the phosphorylation of LKB-1 (Ser 428) in gastrocnemius skeletal muscle. Values are mean ± SD (n = 6 per group). DMSO = control. CD 100 = curcumin 100 mg/kg-BW/day in DMSO. *: significantly different from the DMSO without exercise group (P < 0.05). #: significantly different from the DMSO + exercise group (P < 0.05). †: significantly different from the curcumin without exercise group (P < 0.05). ‡: significantly different from the curcumin with exercise group.

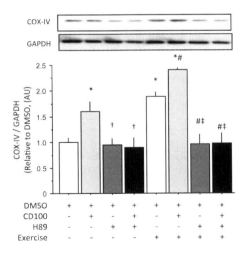

Figure 3. PKA inhibitor H89 abolished the effect of curcumin treatment on COX-IV expression in gastrocnemius skeletal muscle. Values are mean ± SD (n = 6 per group). DMSO = control without exercise. CD 100 = curcumin 100 mg/kg-BW/day in DMSO. *: significantly different from the DMSO without exercise group (P < 0.05). #: significantly different from the DMSO + exercise group (P < 0.05). †: significantly different from the curcumin without exercise group (P < 0.05). ‡: significantly different from the curcumin + exercise group.

This result also indicated that curcumin treatment alone increased COX-IV expression, and the additive effect of curcumin in combination with exercise increased COX-IV expression. Furthermore, the PKA inhibitor (H89) abolished this effect.

4 DISCUSSION

cAMP serves as a second messenger within a cell. Our previous result indicated that curcumin increases the cAMP levels in gastrocnemius skeletal muscle (Ray Hamidie et al., 2015). A previous study has shown that curcumin inhibits PDE in endothelial cells (Abusnina et al., 2009) and pancreatic β cells (Rouse et al., 2014). Furthermore, Another polyphenol (resveratrol) increases cAMP levels, not by increasing cAMP production, but by inhibiting cAMP PDEs, which hydrolyze cAMP to AMP (Chung et al., 2012). This result shows that the polyphenol more preferably inhibits PDE to increase the cAMP levels. Furthermore, we are also interested to examine the effect of curcumin treatment in combination with exercise. As predicted, our result indicated that the additive effect of curcumin treatment and in combination with exercise increased the cAMP levels in skeletal muscle (Fig. 1). Exercise increases the levels of glucagon and catecholamines, which bind to their receptors and increase cAMP production by activating adenylate cyclase (Chung et al., 2012). Based on this evidence, we speculated that the magnitude of cAMP elevation influenced by the additive effect of curcumin treatment in combination with exercise is associated with the increase in adenyl cyclase activity and the inhibition of PDE.

The most common downstream effector of cAMP is Protein Kinase A (PKA) (Tasken and Aandahl, 2004). When a molecule of PKA binds to four molecules of cAMP, the PKA molecule releases two subunits to induce enzyme activity on target proteins (Skalhegg and Tasken, 2000), including phosphorylation of LKB-1 and CREB. Indeed, our result indicated that curcumin administered alone as well as the additive effect of curcumin in combination with exercise have the ability to increase the phosphorylation of LKB-1 and CREB in gastrocnemius skeletal muscle (Figs. 1 and 2). Our result is consistent with that of the previous study that examined the effect curcumin in diabetic rats, showing that curcumin increases the phosphorylation of LKB-1 in the skeletal muscle of rats (Na et al., 2011) and increases the phosphorylation of CREB in the cerebral cortex of rats (Kumar et al., 2010). LKB-1 is the upstream target from AMPK that can act as an energy sensor of the cell, and works as a key regulator of mitochondrial biogenesis. However, it has been reported that PKA is associated with the increased phosphorylation of LKB-1 (Collins et al., 2000). Since PKA activity is dependent on cAMP, we suggested that cAMP induced the phosphorylation of LKB-1 in skeletal muscle. In contrast, CREB localized in the nucleus plays an important role in the regulation of mitochondrial biogenesis due to its ability to increase PGC-1α as the master regulation of mitochondrial biogenesis (Wu et al., 2006b). To verify this hypothesis, we examined the effect of H89 (PKA inhibitor) in order to determine the effect of curcumin treatment on the regulation of mitochondrial biogenesis in skeletal muscle. Indeed, our result indicated that H89 abolished the effect of curcumin treatment on the phosphorylation of LKB-1 and CREB. Furthermore, the additive effect of curcumin treatment in combination with exercise increased the phosphorylation of LKB-1 and CREB, but H89 abolished this effect. These findings indicated that PKA plays an important role in the regulation of mitochondrial biogenesis in skeletal muscle via the LKB-1 and CREB pathway. Our result is also consistent with that of another previous study suggesting that resveratrol-stimulated AMPK activity is abolished in LKB-1-deficient neurons (Dasgupta and Milbrandt, 2007). Indeed, our result also indicated that H89 had the ability to abolish the effect of curcumin treatment on the phosphorylation of LKB-1 (Fig. 1). In the other pathway, CREB has the ability to increase mitochondrial biogenesis in muscle cells through PGC-1α (Wu et al., 2006a). Furthermore, H89 inhibits the effect of CREB. The result of that study is consistent with ours which showed that H89 inhibits the effect of exercise on the phosphorylation of CREB in the skeletal muscle of rodents (Fig. 2). The result of a study conducted in human skeletal muscle after 3 weeks of exercise indicated that CREB phosphorylation at Ser[133] was higher in the skeletal muscle of both legs after training. Evidently, our result indicated that the PKA inhibitor H89 abolished the effect of curcumin treatment alone and in combination with exercise on the phosphorylation of LKB-1 and CREB that is associated with mitochondrial biogenesis in skeletal muscle (Figs. 2 and 3). Based on the above result, we suggested that the PKA inhibitor H89 inhibits the phosphorylation of LKB-1 and CREB associated with the effect of curcumin administered alone and in combination with exercise in skeletal muscle.

However, our result indicated that H89 abolished the effect of curcumin treatment alone and in combination with exercise on the protein expression of COX-IV (Fig. 3). Based on this result, we speculated that deacetylation of PGC-1α is not necessarily associated with the effect of curcumin on the regulation of mitochondrial biogenesis in skeletal muscle. On the one hand, our experimental result indicated that 30 days of endurance exercise (eTR) increased the deacetylation of PGC-1α for regulating mitochondrial biogenesis (Ray Hamidie et al., 2015), and on the other hand, it was found that 3 days of exercise did not sufficiently decrease the acetylation of PGC-α. Another reason is that this condition occurs due to curcumin treatment to increase mitochondrial biogenesis through different pathway beside the LKB-1 and CREB pathway. Future study is necessary to describe the associated effect of curcumin on PGC-1α for regulating mitochondria.

5 CONCLUSION

In conclusion, curcumin treatment together with exercise increased the downstream target of PKA, including the phosphorylation of LKB-1, CREB, deacetylation of PGC-1α, and COX-IV expression in the skeletal muscle of rodents. However, the PKA inhibitor H89 was able to abolish the effect of curcumin treatment together with exercise on cAMP levels and the downstream target of PKA including the phosphorylation of LKB-1, CREB, and COX-IV expression which are associated with the regulation of mitochondrial biogenesis in skeletal muscle. Collectively, our data suggest that inhibition of cAMP synthesis abolishes the effect of curcumin administration in the skeletal muscle of rodents.

REFERENCES

Abusnina, A., Keravis, T. & Lugnier, C. 2009. D020 The polyphenol curcumin inhibits in vitro angiogenesis and cyclic nucleotide phosphodiesterases (PDEs) activities similarly to PDE inhibitors. *Archives of Cardiovascular Diseases*, 102, Supplement 1, S43.

Blough, E., Dineen, B. & Esser, K. 1999. Extraction of nuclear proteins from striated muscle tissue. *Biotechniques*, 26, 202–4, 206.

Chowanadisai, W., Bauerly, K.A., Tchaparian, E., Wong, A., Cortopassi, G.A. & Rucker, R.B. 2010. Pyrroloquinoline quinone stimulates mitochondrial biogenesis through cAMP response element-binding protein phosphorylation and increased PGC-1alpha expression. *J Biol Chem*, 285, 142–52.

Chung, J.H., Manganiello, V. & Dyck, J.R.B. 2012. Resveratrol as a calorie restriction mimetic: therapeutic implications. *Trends in Cell Biology*, 22, 546–554.

Collins, S.P., Reoma, J.L., Gamm, D.M. & Uhler, M.D. 2000. LKB1, a novel serine/threonine protein kinase and potential tumour suppressor, is phosphorylated by cAMP-dependent protein kinase (PKA) and prenylated in vivo. *Biochem J*, 3, 673–80.

Dasgupta, B. & Milbrandt, J. 2007. Resveratrol stimulates AMP kinase activity in neurons. *Proceedings of the National Academy of Sciences*, 104, 7217–7222.

Furuichi, Y., Sugiura, T., Kato, Y., Shimada, Y. & Masuda, K. 2010. OCTN2 is associated with carnitine transport capacity of rat skeletal muscles. *Acta Physiol*, 200, 57–64.

Kawanaka, K., Tabata, I., Katsuta, S. & Higuchi, M. 1997. Changes in insulin-stimulated glucose transport and GLUT-4 protein in rat skeletal muscle after training. *J Appl Physiol*, 83, 2043–2047.

Kumar, T.P., Antony, S., Giressh, G., Geoge, N. & Paulose, C.S. 2010. Curcumin modulates dopaminergic receptor, CREB and phospholipase C gene expression in the cerebral cortex and cerebellum of streptozotocin induced diabetic rats. *J Biomed Sci*, 17, 1423–0127.

Na, L.X., Zhang, Y.L., Li, Y., Liu, L.Y., Li, R., Kong, T. & Sun, C.H. 2011. Curcumin improves insulin resistance in skeletal muscle of rats. *Nutr Metab Cardiovasc Dis*, 21, 526–33.

Nourshahi, M., Gholamali, M., Salehpour, M., Damirchi, A. & Babaei, P. 2012. *Mitochondrial Biogenesis in Skeletal Muscle: Exercise and Aging*, Croatia: intech.

Park, S.J., Ahmad, F., Philp, A., Baar, K., Williams, T., Luo, H., Ke, H., Rehmann, H., Taussig, R., Brown, Alexandra L., Kim, Myung K., Beaven, Michael A., Burgin, Alex B., Manganiello, V. & Chung, Ja H. 2012. Resveratrol Ameliorates Aging-Related Metabolic Phenotypes by Inhibiting cAMP Phosphodiesterases. *Cell*, 148, 421–433.

Ray H.R.D., Yamada, T., Ishizawa, R., Saito, Y. & Masuda, K. 2015. Curcumin treatment enhances the effect of exercise on mitochondrial biogenesis in skeletal muscle by increasing cAMP levels. *Metabolism*, 64, 1334–47.

Rouse, M., Younes, A. & Egan, J.M. 2014. Resveratrol and curcumin enhance pancreatic beta-cell function by inhibiting phosphodiesterase activity. *J Endocrinol*, 223, 107–17.

Skalhegg, B.S. & Tasken, K. 2000. Specificity in the cAMP/PKA signaling pathway. Differential expression, regulation, and subcellular localization of subunits of PKA. *Front Biosci*, 1, D678–93.

Tasken, K. & Aandahl, E.M. 2004. Localized effects of cAMP mediated by distinct routes of protein kinase A. *Physiol Rev*, 84, 137–67.

Than, T.A., Lou, H., Ji, C., Win, S. & Kaplowitz, N. 2011. Role of cAMP-responsive element-binding protein (CREB)-regulated transcription coactivator 3 (CRTC3) in the initiation of mitochondrial biogenesis and stress response in liver cells. *J Biol Chem*, 286, 22047–54.

Veeranki, S., Hwang, S.H., Sun, T., Kim, B. & Kim, L. 2011. LKB1 regulates development and the stress response in Dictyostelium. *Developmental Biology*, 360, 351–357.

Wu, Z., Huang, X., Feng, Y., Handschin, C., Feng, Y., Gullicksen, P.S., Bare, O., Labow, M., Spiegelman, B. & Stevenson, S.C. 2006a. Transducer of regulated CREB-binding proteins (TORCs) induce PGC-1α transcription and mitochondrial biogenesis in muscle cells. *Proceedings of the National Academy of Sciences*, 103, 14379–14384.

Wu, Z., Huang, X., Feng, Y., Handschin, C., Gullicksen, P.S., Bare, O., Labow, M., Spiegelman, B. & Stevenson, S.C. 2006b. Transducer of regulated CREB-binding proteins (TORCs) induce PGC-1alpha transcription and mitochondrial biogenesis in muscle cells. *Proc Natl Acad Sci U S A*, 103, 14379–84.

Yoboue, E.D., Augier, E., Galinier, A., Blamcard, C., Pinson, B., Casteilla, L., Rigoulet, M. & Devin, A. 2012. cAMP-induced mitochondrial compartment biogenesis: role of glutathione redox state. *J Biol Chem*, 287, 14569–78.

Effect of Monosodium Glutamate (MSG) on spatial memory in rats (*Rattus norvegicus*)

R. Razali
Faculty of Medicine, Universitas Syiah Kuala, Aceh, Indonesia

S. Redjeki & A.A. Jusuf
Faculty of Medicine, Universitas Indonesia, Jakarta, Indonesia

ABSTRACT: Monosodium Glutamate (MSG) is a commonly used flavor enhancer in food. Several studies have shown that high doses of MSG act as a neurotoxic or excitotoxic agent for cells in the central nervous system. This study aimed to determine the effect of MSG on spatial memory function. A total of 25 albino male rats of Sprague–Dawley strain (aged 8–10 weeks, weighing 150–200 g) were divided into five experimental groups (two control groups and three treatment groups that received 2 mg/g, 4 mg/g, or 6 mg/g oral MSG, respectively, for 30 days). Spatial memory test was performed using water-E maze before MSG administration and every week (5 times). After the last administration of MSG, all rats were killed. The data were analyzed by one-way ANOVA followed by the Least Significant Difference (LSD) *post hoc* test. The water-E maze test showed an increase in the number of errors made by the treatment group that received 4 mg/g and 6 mg/g of MSG, as well as an increase in time span to complete a memory test when compared with the control group after consumption of MSG for 30 days. In conclusion, this study showed that high MSG doses lead to a reduction in spatial memory function in rats.

Keywords: Monosodium Glutamate (MSG), flavor enhancer, spatial memory, rat, water e-maze

1 INTRODUCTION

Monosodium Glutamate (MSG) is a chemical commonly used as a flavor enhancer in food. MSG has been used in many countries including Indonesia, and has become a research topic in toxicopharmacology (Sukmaningsih et al., 2011; Rangan & Barceloux, 2009). MSG, when dissolved in water or saliva, dissociates into free salt and becomes the anion of glutamate (glutamic acid) (Sukmaningsih et al., 2011). Glutamate is also produced by the human body bound with other amino acids that make up the protein structure as well as those produced by neurons as a neurotransmitter. Glutamate is the major excitatory transmitter in the central nervous system of mammals. Glutamate is abundant in the cerebral cortex, hippocampal gyrus dentatus, and striatum, indicating that it plays an important role in cognitive functions including learning and memory (Setiawati, 2008).

However, the presence of excessive glutamate can lead to overstimulation of receptors that leads to irreversible cell damage or even cell death (Murrah, 2011). Various studies in the early life period have shown that administration of high concentrations of MSG may act as a neurotoxic or excitotoxic agent. Furthermore, it causes damage to the cells in the central nervous system, resulting in various abnormal histological patterns of the cerebral cortex and hippocampus, cerebral cortex layer depletion, damage to neurons in the primary sensory area of neonatal rat cerebral cortex, damage to the paraventricular nucleus of the hypothalamus arcuatus, and the nucleus of neonates whose parents are fed with MSG. MSG has the ability to penetrate the placenta–blood barrier and the brain–blood barrier (Cekic et al., 2005; Zarate et al., 2001; Yuliana, 2012; Briliantina, 2012).

Based on previous studies and high rates of consumption of Monosodium Glutamate (MSG) in the community, we determined whether MSG affects nerve cell function in the hippocampus and produces negative effects on the nervous system when administered in excess.

2 MATERIALS AND METHODS

This study was approved by Health Research Ethics Committee, Faculty of Medicine Universitas Indonesia/Cipto Mangunkusumo Hospital. This in vivo experimental study used 25 male rats of Sprague–Dawley strain, 8–10 weeks old, with the weight of 150–200 g. We used MSG in the form of

Sodium L—glutamate monohydrate ($C_5H_8NNaO_4$) produced by Merck, Germany. Spatial memory examination was performed by the water-E maze test.

Experimental animals were placed 2 weeks install with stable conditions for acclimation. The study group consisted of five groups: pure control group (K1), solvent control group (K2), MSG treatment group of 2 mg/g BW/day (P1), MSG treatment group of 4 mg/g BW/day (P2), and MSG treatment group of 6 mg/g BW/day (P3).

The spatial memory test using water-E maze was performed at the beginning before treatment, then at the end of every week. Each animal test consisted of three repetitions performed one after another with no break between them. We used the stair as a platform. Both the time required for each animal to reach the target (stair), and the number of errors made on each trial were recorded.

Data were analyzed using parametric statistical test: one-way ANOVA, repeated ANOVA, and subsequent test with 95% confidence intervals. We stated $p < 0.05$ as significant. The Least Significant Difference (LSD) method was used as a *post hoc* test.

3 RESULTS

We found a significant difference in memory test errors between control and MSG treatment groups of 4 mg/g BW and 6 mg/g BW (Table 1). There was a tendency of increase in errors shown after consumption of MSG for four weeks, while errors in control group tend to decrease.

A significant difference was found in the time span between the control group and all MSG treatment groups. It was shown (Table 2) that the time span to complete the memory test was getting shorter every week in all groups. The control group was shown to have shorter average time to complete the test.

4 DISCUSSION

As shown in the present study, MSG has an influence on spatial memory in rats after 4 weeks of use. MSG decreases the ability of spatial memory assessed by an increasing number of mistakes as well as having a long duration of time required to complete the memory test. As errors were increased, it showed that experimental animals are not able to remember exactly the route that has to be followed to quickly reach the target in the form of a ladder to remove themselves from the water-E maze.

It is known that the hippocampus is an area that plays an important role in learning and memory, especially spatial memory (Collison et al., 2010; Frieder & Grimm, 1984; Carlson, 2013). Glutamate itself plays an important role in learning and

Table 1. Average number of errors in memory test.

Memory test	Average number of errors (times/test)				
	K1	K2	P1	P2	P3
0	1.00 ± 0.30	1.26 ± 0.29	1.32 ± 0.25	1.00 ± 0.30	1.00 ± 0.30
1	1.08 ± 0.62	1.20 ± 0.54	1.00 ± 0.30	1.20 ± 0.37	1.20 ± 0.58
2	0.66 ± 0.41	0.60 ± 0.30	0.80 ± 0.37	0.94 ± 0.25	0.94 ± 0.26
3	0.60 ± 0.30	0.68 ± 0.25	1.06 ± 0.25	0.68 ± 0.25	0.80 ± 0.37
4	0.60 ± 0.30	0.68 ± 0.25	1.00 ± 0.30	1.20 ± 0.37*	1.28 ± 0.54*

Note: Data expressed as mean ± SD.

Table 2. Average time span in the memory test.

Memory test	Average time span (seconds)				
	K1	K2	P1	P2	P3
0	39.40 ± 7.92	35.46 ± 15.52	39.80 ± 8.54	37.58 ± 10.61	45.54 ± 14.11
1	31.26 ± 17.17	29.26 ± 10.87	38.80 ± 12.52	29.74 ± 10.47	36.00 ± 11.58
2	30.32 ± 8.77	23.48 ± 9.19	23.40 ± 5.50	30.86 ± 5.00	29.40 ± 6.23
3	27.14 ± 4.92	24.54 ± 4.37	29.34 ± 3.37	27.26 ± 8.42	23.00 ± 4.36
4	15.94 ± 2.14	13.27 ± 2.53	20.52 ± 2.60*	20.86 ± 2.96*	19.58 ± 4.05*

Note: Data expressed as mean ± SD.

memory especially in the formation of Long-Term Potentiation (LTP), which allegedly is a process that is responsible for synaptic changes during learning process taking place through the activation of the ion channel receptors, the NMDA receptor t in the hippocampus (Collison et al., 2010; Frieder & Grimm, 1984; Scheetz & Constantine, 1994).

However, if there is an excess of glutamate levels more than normal conditions, resulting in the accumulation of glutamate in the number of abundance in the synaptic cleft, then this will only be excitotoxic to nerve tissues (Ardyanto, 2004). This accumulation will lead to overstimulation of glutamate receptors, especially the NMDA receptor which activates multiple ways that will eventually cause damages and even death of neurons. These damages to the structure of neurons certainly affect the function of neural nerve tissues (Murrah, 2011; Purves et al., 2004; Siegel & Sapru, 2011).

Research of Onaolapo OJ et al (2012) showed that the use of MSG at doses of 1 and 1.5 mg/kg for 14 days induces a tendency of decreased ability of spatial learning and memory in mice in the treatment group compared with the control group tested with the Y-maze although there were no statistically significant differences (Onaolapo et al., 2012). Research by Narayanan et al (2010) also showed that MSG significantly affected the behavioral performance in young adult rats which also increased anxiety and memory retention that is assessed through passive avoidance test (Narayan et al., 2010). Yu L et al. (2011) also proved that the administration of MSG reduced the learning ability and memory retention ability that was assessed using the Y-maze.

Pavlovic et al (2007) mentioned that the death of neurons that occurs as a result of the use of MSG is most likely due to the effects of glutamate excitotoxicity triggered by elevated levels of Ca^{2+} in the cytosol and is mediated through the process of necrosis and apoptosis mechanism of neuronal excitotoxicity, which occurred from a variety of factors (Pavlovic, 2007). The existence of excessive stimulation of glutamate receptors will be able to initiate a variety of potential cascade to induce cell damage and death. NMDA receptors activated by glutamate molecules cause an influx of calcium ions (Ca^{2+}) in bulk, accompanied by sodium ions (Na^+) through AMPA receptors. Moreover, it can also activate metabotropic glutamate receptors causing increased release of Ca^{2+} ions stored in the endoplasmic reticulum (Erdmann et al., 2006). This condition can trigger a variety of pathways that will eventually cause damage to the synapse and cell death, which can be either apoptosis or necrosis.

In this study, the mechanisms described above may play a role in nerve cell damage in specific brain areas associated with spatial memory function, especially in the hippocampus area. The hippocampus also has a role in navigation capabilities, so that if the hippocampus is not functioning properly, then the individual cannot remember the path/route to his destination or path/route to the place it belongs (Bear et al., 2007). Treatment with MSG affects the appearance or animal ability in completing the memory test after administration of MSG with a duration of 4 weeks.

5 CONCLUSION

The result of this study showed that high MSG doses lead to a reduction in spatial memory function in rats after consumption of MSG for four weeks.

REFERENCES

Ardyanto TD. MSG dan kesehatan: sejarah, efek dan kontroversinya. INOVASI. 2004; 1(XVI): 52–6.
Bear MF, Connors BW, Paradiso MA. Neuroscience exploring the brain. Philadelphia: Lippincott Williams & Wilkins; 2007.
Brilliantina L. Pengaruh pemberian monosodium glutamate pada induk tikus hamil terhadap berat badan dan perkembangan otak anaknya pada usia 7 dan 14 hari [Tesis]. Jakarta: FKUI; 2012.
Carlson NR. Physiology of behavior. 11th edition. Boston: Pearson Education; 2013.
Cekic S, Filipovic M, Pavlovic V, Ciric M, Nesic M, Jovic Z, et al. Histopathologic changes at the hypothalamic, adrenal and thymic nucleus arcuatus in rats treated with monosodium glutamate. Acta Medica Medianae. 2005; 44: 35–42.
Collison KS, Makhoul NJ, Inglis A, Al-Johi M, Zaidi MZ, Maqbool Z, et al. Dietary trans-fat combined with monosodium glutamate induces dyslipidemia and impairs spatial memory. Physiol Behav. 2010; 99(3): 334–42.
Danbolt NC. Glutamate as a neurotransmitter-an overview. Prog Neurobiol. 2001; 65: 1–105.
Dobrek L, Thor P. Glutamate NMDA receptors in pathophysiology and pharmacotherapy of selected nervous system diseases. Postepy Hig Med Dosw (online). 2011; 65: 338–46.
Erdmann NB, Whitney NP, Zheng J. Potentiation of excitotoxicity in HIV-1 associated dementia and the significance of glutaminase. Clin Neurosci Res. Elsevier. 2006; 6(5): 315–28.
Frieder B, Grimm VE. Prenatal monosodium glutamate (MSG) treatment given through the mother's diet causes behavioral deficits in rat offspring. Intern J Neurosci. 1984; 23: 117–126.
Michaelis EK. Molecular biology of glutamate receptors in the central nervous system and their role in excitotoxicity, oxidative stress and aging. Prog Neurobiol. 1998; 54: 369–415.
Murrah JD. The dangers of MSG. JD Murrah on hubpages. Web. 2011.

Narayanan SN, Kumar RS, Paval J, Nayak S. Effect of ascorbic acid on the monosodium glutamate-induced neurobehavioral changes in periadolescent rats. Bratisl lek Listy. 2010; 111(5): 247–52.

Onaolapo OJ, Onaolapo AY, Mosaku TJ, Akanji OO, Abiodun OR. Elevated plus maze and Y-maze behavioral effects of subchronic, oral low dose monosodium glutamate in Swiss albino mice. IOSR Journal of Pharmacy and Biological Sciences (IOSR-JPBS). 2012; 3(4): 21–7.

Pavlovic V, Cekic S, Kocic G, Sokolovic D, Zivkovic V. Effect of Monosodium glutamate on apoptosis and Bcl-2/Bax protein level in rat thymocyte culture. Physiol Res. 2007; 56: 619–26.

Purves D, Augustine GJ, Fitzpatrick D, Hall WC, LaMantia AS, McNamara JO, et al. Neuroscience. 3rd edition. Massachusetts USA: Sinauer Associates, Inc; 2004.

Rangan C, Barceloux DG. Food additives and sensitivities. Disease-a-month. 2009; 55(5): 292–311.

Scheetz AJ, Constantine PM. Modulation of NMDA receptor function: implication for vertebrate neuronal development. FASEB J. 1994; 8: 745–52.

Sengul G, Suleyman C, Cakir M, Coban MK, Saruhan F, Hacimuftuoglu A. Neuroprotective effect of ACE inhibitors in glutamate-induced neurotoxicity: rat neuron culture study. Turkish Neurosurgery. 2011; 21(3): 367–71.

Setiawati FSN. Dampak penggunaan monosodium glutamat (MSG) terhadap kesehatan lingkungan. ORBITH. 2008; 4(3): 453–9.

Siegel A, Sapru HN. Essential Neuroscience. 2nd edition. Philadelphia: Lippincott Williams & Wilkins; 2011.

Smith QR. Glutamate and glutamine in the brain: transport of glutamate and other amino acids at the blood-brain barrier. American Society for Nutritional Sciences. J Nutr. 2000; 130: 1016–22.

Sukmaningsih AA, Ermayanti IGA, Wiratmini NI, Sudatri NW. Gangguan spermatogenesis setelah pemberian monosodium glutamat pada mencit (Mus musculus L). Jurnal Biologi XV. 2011; 2: 49–52.

Yu L, Ma J, Ma R, Zhang Y, Zhang X, Yu T. Repair of excitotoxic neuronal damage mediated by neural stem cell lysates in adult mice. J Cell Sci Ther. 2011; 2(3): 1–5.

Yuliana I. Gambaran histologis serebrum neonatus tikus Sprague dawley yang induknya terpapar monosodium L-glutamat selama gestasi [Tesis]. Jakarta: FKUI; 2012.

Zarate CB, Huizar SVR, Contreras AM, Velasco AF, Borunda JA. Changes in NMDA-receptor gene expression are associated with neurotoxicity induced neonatally by glutamate in the rat brain. Neurochem Int. Elsevier Science Ltd. 2001; 39: 1–10.

Mycobacterium tuberculosis load and rifampicin concentration as risk factors of sputum conversion failure

E. Rohmawaty, H.S. Sastramihardja & R. Ruslami
Department of Pharmacology and Therapy, Faculty of Medicine, Universitas Padjadjaran, Bandung, West Java, Indonesia

M.N. Shahib
Department of Biochemistry and Molecular Biology, Faculty of Medicine, Universitas Padjadjaran, Bandung, West Java, Indonesia

ABSTRACT: Pulmonary TB is a chronic infectious disease that has become a global health burden. The aim of this study is to understand the role of bacterial load and plasma rifampicin concentration as risk factors of sputum conversion failure in pulmonary TB. This is a quantitative study with a cross-sectional design conducted on 102 subjects. The study was conducted at the RSP Faculty of Medicine, Universitas Padjadjaran. Plasma rifampicin concentration was measured using UPLC, sputum conversion was examined by Ziehl–Neelsen staining, and statistical analysis was performed using SPSS version 20. The results indicate that the baseline *Mycobacterium tuberculosis* (M.tb) load was significantly different between the groups (p = 0.04), and high M.tb load was the risk factor of sputum conversion failure (POR = 2.97 CI 95% = 1.23–7.15, p = 0.015). In conclusion, M.tb load was found to be the risk factor of sputum conversion failure, not the plasma rifampicin concentration.

1 INTRODUCTION

Pulmonary TB is a chronic infectious disease caused by *Mycobacterium tuberculosis* (M.tb), which has become a global health burden. M.tb typically affects the lungs (pulmonary TB) as well as any other organ of the body (extra-pulmonary TB). The disease spreads when individuals with pulmonary TB expel bacteria into the air, for example, by coughing. Overall, a relatively small proportion (5–15%) of the estimated 2–3 billion people infected with M.tb will develop the TB disease during their lifetime. In 2015, an estimated number of 10.4 million new (incident) TB cases were reported worldwide, of which 5.9 million (56%) were among men, 3.5 million (34%) among women, and 1.0 million (10%) among children. Six countries accounted for 60% of new cases: India, Indonesia, China, Nigeria, Pakistan, and South Africa. There were an estimated 1.4 million TB deaths in 2015. TB remained one of the top 10 causes of death worldwide in 2015. In the same year, 6.1 million new TB cases were reported to national authorities and the WHO. One of the diagnostic tests for TB is sputum smear microscopy. This diagnostic technique for tuberculosis was developed more than 100 years ago. Sputum samples are examined under a microscope to determine whether bacteria are present. In the current case definitions recommended by the WHO, one positive result is required for a diagnosis of smear-positive pulmonary TB (WHO 2016).

Sputum smear microscopy has been the primary method for the diagnosis of pulmonary tuberculosis in low- and middle-income countries, where about 95 per cent of TB cases and 98 per cent of TB deaths occur. It is a simple, rapid, and inexpensive technique that is highly specific in areas with a very high prevalence of tuberculosis. It also identifies the most infectious patients and is widely applicable in various populations with different socio-economic levels. Hence, it has been an integral part of the global strategy for TB control. However, sputum smear microscopy has significant limitations in its performance. The sensitivity is grossly compromised when the bacterial load is less than 10,000 organisms/ml sputum sample (Desikan 2013). A previous study has revealed that bacterial load is association with sputum conversion failure at 2 months after initiation of treatment. Some studies have reported that one of the risks of a persistent positive smear at 2 months was greater in patients with a bacterial load >3+ (Mota et al. 2012). Without appropriate treatment, the death rate from TB is high. Studies of the natural history of the TB disease in the absence of treatment with anti-TB drugs (which were conducted before drug treatments became available) have found that about 70% of people with sputum smear-positive pulmonary TB died within 10 years, as did about 20% of people with culture-positive (but smear-negative) pulmonary TB. Effective drug treatments were first developed in the 1940s.

The currently recommended treatment for new cases of drug-susceptible TB is a 6-month regimen of four first-line drugs: isoniazid, rifampicin, ethambutol, and pyrazinamide (WHO 2016).

Rifampin is considered to be the cornerstone in the current treatment of TB. Rifampin inhibits the β-subunit of the RNA polymerase, a multi-subunit enzyme that transcribes bacterial RNA. Rifampin exhibits concentration-dependent activity that correlates best with the AUC/MIC ratio, as has been shown in a mouse model. The AUC is an important parameter for concentration-dependent killers such as the rifamycins. It indicates total exposure to the drug over a certain time period. Results from an efficacy study in mice predicted one-third reduction in the TB treatment duration when the rifampin dose was increased by 50%. Only a few data are available on the efficacy of regimens based on a higher dose of rifampin in human subjects. A short regimen of a high dose of rifampin (1,200 mg daily or every other day) combined with a high dose of isoniazid (900 mg) and streptomycin (1,000 mg) daily yielded almost 100% sputum culture conversion after 3 months (Boogaard et al. 2009).

Generally, first-line treatment of drug-susceptible tuberculosis (TB) is highly effective. However, a number of patients do not respond adequately to the treatment, and develop drug resistance or experience a relapse of TB after completion of treatment. Inadequate exposure to anti-TB drugs may constitute one of the factors underlying a suboptimal treatment response. Some studies have reported associations between low concentrations of anti-TB drugs and poor treatment responses, although this has not been found in other studies (Burhan et al 2013).

The role of bacterial load and the efficacy of rifampicin as a primary drug of TB should be explored to understand the disease and for therapeutic improvement. The aim of this study is to understand the role of bacterial load and plasma rifampicin concentration as risk factors of sputum conversion failure in pulmonary TB.

2 MATERIALS AND METHODS

2.1 *Study design and recruitment of subjects*

We conducted a case control study among TB patients who visited the Teaching Hospital (RSP) Faculty of Medicine, Universitas Padjadjaran, Bandung, for treatment between April 2014 and April 2015. Patients were included in the study if they had a positive Acid-Fast Bacilli (AFB) smear, were undergoing treatment with anti-TB drugs according to the direct observed treatment short course (DOT) strategy, and gave informed consent. Patients were excluded from the study if they were pregnant, below 18 or above 65 years of age, had chronic diarrhea, and impaired liver or renal function. Drug dosing was according to WHO-recommended weight bands. The anti-TB drugs were administered in fixed dose combinations that were approved by the Indonesian national program based on bio-equivalence studies. Anti-TB drugs were taken without food.

2.2 *Blood sampling, bioanalysis, and pharmacokinetic data analysis*

Patients refrained from food at least 8 h before drug intake at the pharmacokinetic sampling day. Plasma drug concentration collected from blood 2 h after administration of drugs ($C2$ h) was used as the estimated peak plasma concentration (Cmax). Drug sampling took place at 4 weeks after inclusion in the study because of the expected steady state in the pharmacokinetics of TB drugs at that time. Plasma was separated and stored at 80°C immediately until analysis. Plasma rifampicin concentrations were measured with a validated method such as liquid–liquid extraction, followed by ultraperformance liquid chromatography with UV detection (Burhan et al 2013). Sputum smears were examined for AFB by fluorescence microscopy and/or Ziehl–Neelsen staining and graded by standard criteria and equivalence: 1–9 AFB/100 fields (1+); 1–9 AFB/10 fields (2+); 1–9 AFB/field (3+); and >9 AFB/field (4+) (Mota et al 2012). Sputum smears were done at the first subject examination and at the end of the initial treatment phase (at the 8th week) to determine smear conversion. Subjects with conversion failure were categorized as the case group and those with conversion were categorized as the control group.

Statistical analysis was performed using the SPSS version 20.0 software (SPSS Inc., Chicago, Illinois, USA). All probabilities were two-tailed and p values < 0.05 were considered as significant.

3 RESULTS

A total of 102 patients with pulmonary TB were recruited for the study. Sputum examination was performed by the Ziehl–Neelsen method to determine the baseline M.tb load. Plasma rifampicin concentration was measured by using UPLC. After treatment with oral anti-tuberculosis, as recommended by the WHO, for 8 weeks, 72 subjects had negative M.tb sputum smear (conversion group) and 30 subjects had positive M.tb sputum smear (non-conversion group). The mean age of the subjects in the conversion and non-conversion groups was 35.3 (SD 11.5) years and 34.5 (SD 13.6) years, respectively. Of the total subjects, 59.7% in the conversion group and 50% in the non-conversion group were male (Table 1).

The results indicated that baseline M.tb load was significantly different between the conversion

Table 1. Subject characteristics.

Subject characteristics	Conversion (N = 72)	Non-conversion (N = 30)	p
Sex, n (%)			
Male	43 (59.7%)	15 (50%)	0.37*
Age (yrs)			
Mean (SD)	35.3 (11.5)	34.5 (13.6)	0.751**
Baseline M.tb load			
Scanty	11 (15.3%)	1 (3.3%)	0.04
+1	20 (27.8%)	4 (13.3%)	
+2	19 (26.4%)	8 (26.7%)	
+3	22 (30.6%)	17 (56.7%)	
Rifampicin conc., n (%)			
Mean (SD)	8.9 (4.12)	7.9 (4.45)	0.287**

*p value obtained using chi-square test.
**p value obtained using *t*-test.

Table 2. Multivariate analysis of baseline M.tb load and plasma rifampicin concentration.

Variable	Crude OR	CI 95%	P	Adj OR	CI 95%	P
Rifampicin conc.	1.6	(0.68–3.77)	0.28	1.78	(0.64–4.95)	0.27
High baseline M.tb load (+3)	2.97	(1.23–7.15)	0.015	5.08	(1.72–15.04)	0.03

and non-conversion groups (p = 0.04). In the non-conversion group, 56.7% had high M.tb load (+3). However, in the conversion group, 30.6% had high M.tb load (+3). There was no significant difference between the conversion and non-conversion groups in relation to plasma rifampicin concentration (8.9 mg/L vs. 7.9 mg/L, p = 0.287) (Table 1).

Multivariate analysis indicated that only high baseline M.tb load (+3) is the risk factor of sputum conversion failure (POR = 2.97 CI 95% = 1.23–7.15; p = 0.015) (Table 2). Subjects with high baseline M.tb load had 2.97 times risk for sputum conversion failure compared with those with low baseline M.tb load.

4 DISCUSSION

In this study, most of the subjects in the conversion group were male, and in the non-conversion group the number of male and female subjects was equal. TB is more a disease of men than of women, and gender is an important aspect of TB epidemiology. The difference between men and women appears to be larger in Asia, especially in the SEARO region and to a lesser extent in the WPRO region, than in sub-Saharan Africa (AFRO region). It is unclear why male–female differences in tuberculosis prevalence appear to be larger in many Asian surveys than in those from Africa (Borgdorff et al. 2006).

The results indicated that baseline M.tb load was significantly different between the conversion and non-conversion groups (p = 0.04). A previous study has reported that patients with high pre-treatment smear grade (3+/4+) were less likely to have conversion than patients with low pre-treatment grade (1+/2+). Persistent smear positivity was associated with higher pre-treatment grade due to the initially high mycobacterial burden (Mota et al. 2012). Another study in Indonesia has reported that baseline bacterial load (+3) is associated with very low sputum conversion percentage (Triyani et al. 2007). Knowledge of the risk factors associated with delayed conversion helps identify highly infectious patients who require the most medical resources and prolonged respiratory isolation, and necessitates a cautious interpretation of sputum smear results (Mota et al. 2012).

Rifampicin plasma concentration in the conversion and non-conversion groups was 8.9 mg/L and 7.9 mg/L, respectively. This concentration was measured at 2-hour after drug administration. Targeted peak rifampicin concentration was 8 to 24 g/ml after a 600 mg dose. Recommendation of the dose is increased if the peak is less than 6 g/ml (Peloquin 2002). A pilot study in Indonesian patients was performed to compare a higher dose (600 mg) and the standard dose (450 mg or 10 mg/kg, considering the body weight of Indonesian people) of rifampin. It showed that 78% of the patients in the higher-dose arm versus 48% of those in the standard-dose arm achieved a peak (2-hour) plasma rifampin concentration above a reference value of 8 mg/liter (Ruslami et al. 2007).

A 2-hour post-dose sample can be collected to estimate the peak concentration. Given the variability of oral absorption, a single time point may miss the actual peak concentration. Therefore, a second sample, typically 6 hours post-dose, allows one to capture information on the rate and completeness of absorption. It can also provide information regarding the elimination of drugs that have short half-lives, such as isoniazid and rifampicin, provided that absorption is completed by 2 hours post-dose. If the 2-hour and 6-hour values are roughly the same, perhaps somewhat below the expected ranges, or if the 6-hour values are higher than the 2-hour values, then delayed absorption is most likely. Delayed but normal concentrations do not require dose adjustment, but malabsorption, especially Cmax values less than 4 g/ml, should prompt a dose increase. This allows one to take better advantage of the dose response of rifampicin (Peloquin 2002).

A pharmacokinetic study of rifampicin in Peru revealed very low levels of rifampin at 2 h in 68.6% of patients, which is in accordance with data from other studies. Moreover, 36 patients (34.3%) had undetectable levels at 2 h, when a peak concentration is theoretically reached after oral administration. Wilkins et al. previously reported considerable interindividual variability in rifampin pharmacokinetics in South Africa, and suggested that highly variable rates of absorption could significantly impact the Cmax values as slow absorption could lead to low peak plasma concentrations. In this study, 61 patients had higher levels of rifampin at 6 h than at 2 h, which we attributed to a delay in absorption (Mendez et al. 2012).

Slow absorbance could be associated with genetic variations such as human genetic polymorphisms of drug influx and efflux transporter genes (anion-transporting polypeptides [*SLCO1B1* and *SLCO1B3*] and P-glycoprotein [*MDR1*]). A study by Weiner et al. concluded that lower rifampin exposure was associated with the polymorphism of the *SLCO1B1* c.463C_A gene (Weiner et al. 2010).

In this study, rifampicin plasma concentration was not significantly different between the conversion and non-conversion groups (8.9 mg/L vs. 7.9 mg/L, p = 0.287). A previous study on isoniazid, rifampin, and pyrazinamide plasma concentrations in relation to treatment responses in Indonesian pulmonary tuberculosis patients has reported that there was no association between drug concentrations and 8-week culture conversion. This condition can be explained by the fact that it is not the C2 h or Cmax of TB drugs but the total drug exposure (AUC) that is related to the response. C2 h was used as a surrogate of Cmax instead of using a precisely estimated Cmax derived from repeated sampling at different time points. Nevertheless, C2 h correlated well with Cmax and/or AUC in this and other studies. Cmax/MIC or AUC/MIC could be related to the response, but not the Cmax or AUC. However, the MIC for the various drugs was not determined. Another explanation is that the response to rifampin may better correlate with protein-unbound (free) concentrations than with total (protein-unbound plus bound) concentrations. Drug concentrations would ideally be measured at target locations, e.g., in the epithelial lining fluid instead of plasma. In addition to drug-related factors such as patient and bacterial factors, many other factors also determine the response in an individual patient (Burhan et al. 2013).

The main limitations of this study are the small sample size, one point measurement of plasma rifampicin concentration, and non-follow-up of the subjects every week to determine the time of sputum conversion. These limitations could have affected the result of this study, which indicated that plasma rifampicin concentration was not significantly different between the conversion and non-conversion groups, and also that it was not an associated risk factor of sputum conversion failure.

5 CONCLUSION

This study showed that M.tb load was the risk factor of sputum conversion failure in pulmonary TB, not the plasma rifampicin concentration. More attention needs to be given to pulmonary TB patients with high baseline M.tb load because the risk of sputum conversion failure was higher than that in patients with lower baseline M.tb load. Adherence, treatment supervision by families and health workers, and upgrading of nutritional and health status should be maximized to achieve successful treatment and avoid drug resistance.

REFERENCES

Boogaard, J.V.D., Kibiki, G.S., Kisanga, E.R., Boeree, M.J., Aamoutse, R.E. 2009. New drugs against tuberculosis: problem, progress, evaluation of agents in clinical development. *Antimicrobial Agents and Chemotherapy* 53(3):849–62.

Borgdorff, M., Nagelkerke, N., Dye, C., Nunn, P. 2000. Gender and tuberculosis: A comparison of prevalence surveys with notification data to explore sex differences in case detection. *Int J Tuberc Lung Dis* 4:123–32.

Burhan, E., Ruesen, C., Ruslami, R., Ginanjar, A., Mangunnegoro, H., Ascobat, P., Donders, R., et al. 2013. Isoniazid, rifampin, and pyrazinamide plasma concentrations in relation to treatment response in indonesian pulmonary tuberculosis patients. *Antimicrob Agents Chemother* 57:3614.

Desikan, P. 2013. Perspective sputum smear microscopy in tuberculosis: Is it still relevant? *Indian J Med Res* 137:442–44.

Mendez, A.R., Davies, G., Ardrey, A., Jave, O., Romero, S.L.L. 2012. Pharmacokinetics of Rifampin in Peruvian Tuberculosis Patients with and without Comorbid Diabetes or HIV. In: Ward, S.A., Moorea, D.A.J., editors. *Antimicrobial Agents and Chemotherapy* 56(5):2357–63.

Mota, P.C., Carvalho, A., Valente, I., Braga, R., Duarte, R. 2012. Predictors of delayed sputum smear and culture conversion among a Portuguese population with pulmonary tuberculosis. *Rev Port Pneumol* 18(2):72–9.

Peloquin, C. 2002. Therapeutic drug monitoring in the treatment of tuberculosis. *Drugs* 62:2169–83.

Ruslami, R., Nijland, H., Alisjahbana, B., Parwati, I., Crevel V.R., Aarnoutse, R. 2007. Pharmacokinetics and tolerability of a higher rifampicin dose versus the standard dose in pulmonary tuberculosis patients. *Antimicrob Agents Chemother* 51:2546–51.

Triyani, Y., Parwati, I., Sjahid, I., Gunawan, J. 2007. Peralihan (Konversi) Sputum BTA Antara Pemberian Dosis Baku (Standar) Dan Tinggi Rifampicin Pada Pengobatan (Terapi) Anti Tuberkulosis Kelompok (Kategori) I. *Indonesian Journal of Clinical Pathology and Medical Laboratory* 14:1–10.

Weiner, M., Peloquin, C., Burman, W., Luo, C.C., Engle, M., Prihoda, T.J., Mac Kenzie, W.R. et al. 2010. Effects of tuberculosis, race, and human gene *SLCO1B1* polymorphisms on rifampicin concentrations. *Antimicrobial Agents and Chemotherapy* 54(10):4192–200.

WHO. 2016. Global Tuberculosis Report 2016. Geneva, Switzerland: World Health Organization.

Interleukin-22 serum in comedonal acne vulgaris: Proof of inflammation

K. Ruchiatan, R. Hindritiani & E. Sutedja
Faculty of Medicine, Universitas Padjadjaran, Bandung, West Java, Indonesia
Department of Dermatology and Venereology, Faculty of Medicine, Universitas Padjadjaran—Dr Hasan Sadikin Hospital, Bandung, West Java, Indonesia

S. Maulinda
Department of Dermatology and Venereology, Faculty of Medicine, Universitas Padjadjaran—Dr Hasan Sadikin Hospital, Bandung, West Java, Indonesia

ABSTRACT: Inflammation is one of the important processes in the pathogenesis of Acne Vulgaris (AV). Although clinically comedones are non-inflammatory lesions, microscopically the inflammation is active. Induction of immunological responses to *Propionibacterium acnes* (*P. acnes*) is the main important factor in the development of AV. *P. acnes* is capable of inducing T-helper (Th)-17, the main function of which is to produce Interleukin (IL)-22 that acts as a pro-inflammatory cytokine. We compared IL-22 serum levels between the papulopustular type (inflammatory lesions) and the comedonal type (non-inflammatory lesions) in AV patients. Blood samples were collected from 12 patients for each group of papulopustular and comedonal types of AV, and measured the IL-22 serum level using the ELISA technique. The mean IL-22 serum level in the papulopustular group was 18.18 pg/ml, whereas in the comedonal group, it was 16.87 pg/ml ($p > 0.05$). The detection of IL-22 serum in the comedonal group supports the theory that the inflammatory process begins with comedones.

1 INTRODUCTION

Acne Vulgaris (AV) is a common chronic inflammatory disorder of pilosebaceous units (Zanglein et al. 2003). In the USA, it affects 40–50 million individuals every year, with the highest incidence occurring in young adults (85%) at the age of 12–24 years (Zanglein et al. 2003; James et al. 2006).

The pathogenesis of AV is multifactorial and not yet clearly understood (Kim et al. 2002; Pawin et al. 2004). Inflammation is one of the important processes in the pathogenesis of AV (Koreck et al. 2003). It leads AV into more than a skin-deep disease (Ayer et al. 2006).

Induction of immunological responses to *Propionibacterium acnes* (*P. acnes*) is the main important factor in the development of AV (Koreck et al.2003). A recent study has shown that *P. acnes* stimulates inflammation by producing IL-17 from human peripheral blood mononuclear cells (PBMCs), as well as by producing other cytokines such as IL-1β, IL-6, and TGF-β, which induce the differentiation of T CD4[+] cells into Th17 cells (Agak et al. 2014). Th17 produces IL-22 abundantly, which acts as a pro-inflammatory cytokine, which is also expressed together with IL-17 A and IL-17F during its differentiation (Dessinoti et al.2010).

According to the clinical and therapeutic importance, AV is usually classified based on severity and predominant lesion types such as comedonal, papular, pustular, and conglobata (Oschendorf. 2009). Nevertheless, the exact shifting mechanism from non-inflammatory lesions such as comedones to inflammatory lesions such as papulopustules is yet unclear (Saurat. 2015). Although clinically a comedo is a non-inflammatory lesion, microscopically the inflammation has been found to occur (Jeremy et al.1996).

A study on the assay of IL-22 serum levels in AV patients based on the lesion type is still lacking. Therefore, in this study, we proposed that IL-22 has a role in the inflammatory process in AV.

2 MATERIALS AND METHODS

2.1 *Patients*

A total of 24 patients with AV aged above 14 years were selected by consecutive admission from Cosmetic Dermatology Outpatient Clinic, Dr. Hasan Sadikin Hospital. They were classified into two groups, with each group consisting of 12 comedonal type grade 2 or more and 12 papulopustular type grade 2 or more, respectively. The exclusion criteria were: 1) patients who underwent systemic antibiotic (tetracycline, clindamycin, minocycline, doxycycline, and erythromycin) and anti-inflammatory (steroid) therapy within a period of 4 weeks before the start of the study; 2) patients who were suspected or already diagnosed with other inflammatory skin diseases (psoriasis, leprosy, atopic dermatitis, contact allergy) and systemic diseases (diabetes mellitus, asthma, tuberculosis of the lungs, Systemic Lupus Erythematosus (SLE)) based on the history and physical examination.

This study was approved by the Health Research Ethics Committee, Faculty of Medicine, Universitas Padjadjaran—Dr. Hasan Sadikin Hospital, Bandung, Indonesia. Written informed consent was obtained from each participant after oral explanation about the study.

2.2 Classification of AV

We classified the subjects clinically based on the most predominant lesion types and severity using Plewig and Kligman Criteria (1975) based on lesion counts. The comedonal AV grades are as follows: grade 2, 10–25 comedones; grade 3, 26–50 comedones; and grade 4, > 50 comedones. More than 5 papules and pustules should not be counted on one side of the face. The papulopustular AV grades are as follows: grade 2, 10–20 papules and pustules; grade 3, 21–30 papules and pustules; and grade 4, >30 papules and pustules on one side of the face (Oschendorf. 2009).

2.3 Enzyme-linked immunosorbent assay (ELISA) for IL-22

The serum IL-22 levels were measured from the peripheral blood of the subjects using the Human IL-22 Test Kit (*eBioscience®*, Vienna, Austria), according to the manufacturer's instructions. The serum IL-22 levels (pg/ml) in each patient were used for data analysis.

2.4 Statistical analysis

Results are expressed as means ± SD. A *p* value less than 0.05 was considered as statistically significant. The Mann–Whitney analysis was used to determine the differences in serum IL-22 levels between the two groups (comedonal type and papulopustular type).

3 RESULTS

Table 1 provides the characteristics of the subjects enrolled in this study. We enrolled 16 male patients with 10 papulopustular and 6 comedonal types of

Table 1. Characteristics of the patients.

	Papulopustular AV (n = 12)	Comedonal AV (n = 12)	Total
Gender			
Men	10 (83.3%)	6 (50%)	16 (66.7%)
Women	2 (16.7%)	6 (50%)	8 (33.3%)
Age			
15–19 years	3 (25%)	5 (41.7%)	8 (33.3%)
20–24 years	5 (41.7%)	4 (33.3%)	9 (37.5%)
25–29 years	3 (25%)	1 (8.3%)	4 (16.7%)
30–34 years	1 (8.3%)	2 (16.7%)	3 (12.5%)

Table 2. Serum IL-22 levels in the papulopustular and comedonal types of AV.

IL-22 level (pg/ml)	Papulopustular AV (n = 12)	Comedonal AV (n = 12)	p*
Mean (SD)	18.18 (10.96)	16.87 (5.82)	
Median	15.08	15.08	1,000
Range	38.86	21.48	

* p value was obtained using the Mann–Whitney test.

AV and 8 female patients with 2 papulopustular and 6 comedonal types of AV.

Nine subjects (37.5%) were at the age of 20–24 years, and 8 subjects were at the age of 15–19 years.

Table 2 summarizes the result of serum IL-22 levels found in the two groups. The mean serum IL-22 levels in the papulopustular and comedonal types of AV were 18.18 pg/ml and 16.87 pg/ml, respectively, but the difference was not statistically significant (p = 1.000).

4 DISCUSSION

Inflammation that occurs in AV is mainly induced by immunological reactions against *P. acnes* (Koreck et al. 2003). *P. acnes* can trigger inflammatory responses and induce monocytes to secrete several pro-inflammatory cytokines such as IL-17 and IL-22 (Agak et al. 2014).

There was a significant increase in the number of *P. acnes* in patients with AV compared with non-AV patients. Moreover, this bacteria can be found in papules and pustules (inflammatory lesions) and comedones (non-inflammatory lesions) (Shaheen et al. 2011). *P. acnes* was found in larger numbers in inflammatory lesions compared with non-inflammatory lesions, because inflammatory lesions provided a suitable environment and more nutrients for the colonization of *P. acnes* that led to the inflammation process (Shaheen et al. 2011).

Serum IL-17 levels increased in patients with inflammatory AV compared with the normal subjects (Sugisaki et al. 2009). A study for measuring serum IL-22 levels in patients with AV based on severity and predominant lesion types is still lacking.

In this study, there was no difference in the serum IL-22 levels between the papulopustular and comedonal types of AV. These results indicate that inflammation had already occurred in the comedonal type of AV. Comedones are a non-inflammatory lesions clinically, but this lesion has been observed as inflammatory lesions microscopically, firstly by the discovery of the pro-inflammatory cytokine IL-1α in comedonal lesions (Jeremy et al. 1998).

5 CONCLUSION

The results of the study indicated that the serum IL-22 level in the papulopustular type of AV is not different from that in the comedonal type of AV, which strengthens the theory that inflammatory events are already active in comedonal lesions.

REFERENCES

Agak GW, Qin M., Nobe J., Kim MH, Krutzik SR, Tristan GR, et.al. 2014. *Propionibacterium acnes* induces an IL-17 response in acne vulgaris that is regulated by vitamin A and vitamin D. *Journal of Investigation of Dermatology*. 134:366–73.

Ayer J., Burrows N. 2006. Acne: more than skin deep. *Postgraduate medical journal* 82(970):500–6.

James WD, Berge TG, Elston DM. 2006. *Andrews' diseases of the skin*: 231–9. 10th edition. Philadelphia. WB Saunders Company.

Jeremy AHT, Holland DB, Roberts SG, Thomson KF, Cunliffe WJ. 1998. Inflammatory events are involved in acne lesion initiation. *Journal of Investigative Dermatology*. 107(488):20–7.

Kim J., Ochoa MT, Krutzik SR, Takeuchi O., Uetmatsu S., Legaspi AJ, et al. 2002. Activation of tool-like receptor 2 in acne triggers inflammation cytokine responses. *The Journal of Immunology*. 169:1535–41.

Koreck A, Pivarcsi A, Dobozy A, Kemeny L. 2003. The role of innate immunity in the pathogenesis of acne. *Dermatology*. 206(1):96–105.

Oschendorf F. 2009. Acne vulgaris. *CME Dermatology*. 4(1):36–51.

Saurat JH. Strategic target in acne: the comedone switch question. *Dermatology*. 2015;231:105–111

Shaheen B, Gonzales M. 2011. A microbial aetiology: what is the evidence?. British Journal of Dermatology. 165(1):474–85.

Sugisaki H, Yamanaka K, Kakeda M, et al. 2009. Increased interferon-g, interleukin-12p40 and IL-8 production in *Propionibacterium acnes*-treated peripheral blood mononuclear cells from patient with acne vulgaris host response but not bacterial species is the determinant factor of the disease. *Journal of Dermatology Science*. 55(1):47–52.

Zaenglein AL, Graber EM, Thiboutot DM, Strauss JS. 2011. Acne vulgaris and acneiform eruption. In: *Fitzpatrick's dermatology in general medicine*: 690–7028th edition. New York: McGraw Hill.

Advances in Biomolecular Medicine – Hofstra, Koibuchi & Fucharoen (Eds)
© 2017 Taylor & Francis Group, London, ISBN 978-1-138-63177-9

Pharmacokinetic optimization of the treatment of TB meningitis with TB drugs

R. Ruslami
Department of Pharmacology and Therapy, TB-HIV Research Center Faculty of Medicine, Universitas Padjadjaran, West Java, Indonesia

ABSTRACT: Tuberculous Meningitis (TBM) is the most severe form of TB with high mortality and morbidity. The treatment of TBM is not evidence-based, but is similar to the treatment of pulmonary TB. As a key drug in the treatment of TB, rifampicin—a concentration-dependent antibiotic—has a low penetration into the CSF. The current dose of rifampicin is very low and a higher dose of rifampicin might improve the treatment outcome. A series of studies have been conducted to determine whether a higher dose of rifampicin would give a better pharmacokinetic profile, be tolerated by patients, and improve the treatment outcome in TBM patients. This work is the result of collaborative research, funded by different research grants, both national and international.

1 INTRODUCTION

Tuberculous Meningitis (TBM) is the most severe manifestation of TB, leaving up to 50% of patients dead or neurologically disabled (Ganiem et al., 2009). The current treatment of TBM is similar to the treatment of pulmonary TB, although its underlying pathogenesis and organ involved are different from pulmonary TB. Rifampicin is an important drug in the treatment of TB; however, it has a low penetration into the blood–brain barrier (Donald, 2010). There is little evidence about the existence of other TB drugs for the treatment of TBM. Furthermore, the potency of other new anti-TB drugs for TBM has not yet been explored. For example, moxifloxacin, a very potent quinolone for *Mycobacterium tuberculosis* (M.tb) (Shandil et al., 2007), has a good penetration into the CSF (Kanellakopoulou et al., 2008).

With regard to the optimization of the treatment of TBM, there are two ways of approach: non-pharmacological and pharmacological approach. The non-pharmacological approach involves early detection and diagnosis of TBM, or finding a better diagnostic tool. The pharmacological approach involves either finding new drugs that have the potency to kill M. tb and the ability to penetrate into the blood–brain barrier or optimizing the use of available drugs (e.g. rifampicin, isoniazid, pyrazinamide).

2 RIFAMPICIN FOR THE TREATMENT OF TBM

Rifampicin is the cornerstone for the treatment of TB. This drug has been used for almost half a decade. Owing to its limited penetration into the brain, it has been suggested that the current dose of rifampicin may be too low for TBM patients. Rifampicin is a concentration-dependent antimicrobial drug; its higher concentration in plasma as well as in the Cerebrospinal Fluid (CSF) is expected to be more effective.

A higher drug concentration in plasma can be achieved by either using intravenous administration that ensures good bioavailability, or increasing the dose, or both. By understanding the pharmacological property of rifampicin, we hypothesize that a higher dose of rifampicin given intravenously will give a better pharmacokinetic (PK) profile of rifampicin that is still safe and may lead to a better outcome in TBM patients.

3 HIGHER DOSE OF I.V. RIFAMPICIN IN TBM PATIENTS

To test the hypothesis, we conducted an open-label, randomized, phase 2, clinical trial with factorial design (NCT 01158755) in 60 adult TBM patients admitted at Hasan Sadikin General Hospital, Bandung. Eligible subjects were first randomly assigned to receive 600 mg of i.v. rifampicin or 450 mg of oral rifampicin for 14 days. In the second randomization, they were allocated to receive 400 mg and 800 mg of moxifloxacin, or no moxifloxacin (but given ethambutol). INH and pyrazinamide were given to the subjects. The results indicated that increasing the dose of i.v. rifampicin by 1/3 (from 450 mg (about 10 mg/kg in Indonesian patients) to 600 mg (about 13 mg/kg)) led to a three-fold increase in the exposure to rifampicin

Table 1. Pharmacokinetic data of rifampicin (n = 52).

	600 mg, i.v. (n = 26)	450 mg, oral (n = 26)	Ratio of i.v. to oral	p
PLASMA				
AUC0–6 (mg.h/L)	78·7 (71·0–87·3)	26·0 (19·0–35·6)	3·0 (2·2–4·2)	< 0·0001*
Cmax (mg/L)	22·1 (19·9–24·6)	6·3 (4·9–8·3)	3·5 (2·6–4·8)	< 0·0001*
Cmax (≥ 8 mg/L)	26 (100%)	13 (50%)	–	< 0·0001†
Tmax (h; median, range)	2 (1–2)	2 (1–6)	–	0·048‡
CSF				
Cmax (mg/L)§	0·60 (0·46–0·78)	0·21 (0·16–0·27)	2·92 (2·03–4·20)	< 0·0001*

Values are numbers (%) or geometric means (95% CI), unless otherwise indicated. Rifampicin concentrations were measured in samples obtained during the first 3 days of treatment to determine plasma AUC0–6 and CSF Cmax. Assessment of plasma AUC0–24 values was difficult in 41 patients because rifampicin concentrations at C24 were below the limit of quantification. Estimation of C24 based on the last measurable concentration and the elimination rate constant were not possible because the elimination rate for rifampicin could not be assessed reliably with sampling at 2 h, 4 h, and 6 h after dose. For this reason, only the data of AUC0-6 is given. AUC0–6 = area under the time–concentration curve up to 6 h after dose; AUC0–24 = area under the time–concentration curve up to 24 h after dose; Cmax = maximum plasma concentration. Tmax = time to Cmax; CSF = cerebrospinal fluid; C24 = rifampicin concentration at 24 h after dose. *Independent-samples t test after log transformation. †χ2 test. ‡Wilcoxon rank-sum test. §CSF samples were obtained from 25 patients on 600 mg of i.v. rifampicin and 25 patients on 450 mg of oral rifampicin; in one and 16 patients, respectively, rifampicin concentrations in the CSF were below the limit of quantification of the assay (0·26 mg/L) and were set at half the limit of quantification (0·13 mg/L) to enable the comparison of exposures in the CSF (Ruslami et al., 2013).

(AUC) and maximum concentration (C_{max}) of rifampicin in the plasma and CSF (Table 1) (Ruslami et al., 2013).

The distribution of safety/tolerability data, in particular hepatotoxicity data, was equal among the groups. Half of the subjects died within 6 months, and 73% were in the first month of treatment. The cause of death is mainly respiratory failure and neurological deterioration. Surprisingly, mortality was much lower in the high-dose rifampicin arm (adjusted HR 0.42 (95%CI 0.2–0.87, p = 0.0193), as shown in Figure 1. However, there was no additional effect of moxifloxacin use, either standard or high dose, in TBM patients (Ruslami et al., 2013).

These findings indicate that by a simple modification in the use of rifampicin (i.e. increase in the dose and change in the route of administration (intravenous)) in the first 2 weeks of treatment, a better PK profile of rifampicin (AUC and C_{max} in plasma and concentration in the CSF) can be achieved with similar drug-related adverse events, and patient survival can be improved. This is the first evidence of using a higher dose of i.v. rifampicin in TBM patients.

Data obtained from pharmacokinetic/pharmacodynamic (PK/PD) analysis revealed that for patient survival, the minimum target of rifampicin AUC_{0-6h} is ~70 mg.h/L (AUC_{0-24h} of ~116 mg.h/L) and C_{max} is ~22 mg/L (Te Brake et al., 2015).

Furthermore, from these data, we tried to find the optimal doses of rifampicin for pediatric TBM patients using the model-based approach. To achieve the target exposures of rifampicin,

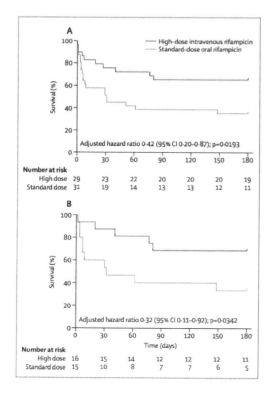

Figure 1. Survival rates after treatment with rifampicin in all the 60 patients (A) and in 31 bacteriologically proven cases of tuberculous meningitis (B) (Ruslami et al., 2013).

children require a daily dose of at least 30 mg/kg oral rifampicin or 15 mg/kg i.v. rifampicin (Savic et al., 2015).

4 HIGHER DOSE OF ORAL RIFAMPICIN

A higher dose of i.v. rifampicin seems promising to achieve a better outcome of TBM treatment. However i.v. rifampicin is not easy; it is invasive, impractical, more expensive than oral drug, and not widely available. For this reason, first, we need to find an alternative for intravenous rifampicin that gives the same effect, i.e. a higher oral dose of rifampicin (i.e. 15 or 20 mg/kg, or even higher) with a similar PK/PD and safety profile to those of i.v. rifampicin. Rifampicin is a widely used drug and tolerated well by patients. In addition, it has been shown that a higher oral dose of rifampicin up to 35 mg/kg is well tolerated in African pulmonary TB patients (Boeree et al., 2015).

Second, we need to find a replacement for i.v. rifampicin with a similar effect. An exploratory PK study (open-label, phase 2, randomized control trial) was performed in 30 adult TBM patients to explore the PK of higher oral dose of rifampicin (750 mg and 900 mg, ~17 and 20 mg/kg in Indonesian patients, respectively) in comparison with i.v. rifampicin (600 mg, ~13 mg/kg) for 14 days. The results indicated that oral rifampicin at doses of 750 mg and 900 mg had less optimal PK profile than i.v. rifampicin at a dose of 600 mg (C_{max} 14.3, 16.2, and 24.7 mg/L, respectively) (Yunivita et al., 2016). Based on the achieved target rates in plasma (AUC_{0-24h} ~116 mg.h/L) (Savic et al., 2015) and rifampicin concentrations in the CSF, follow-up studies have found that even higher doses of rifampicin is needed.

In the next dose-finding study (phase 2b randomized, double-blinded, control trial), we examined the PK of a higher dose of oral rifampicin (twice and thrice that of the standard dose) in 60 adult TBM patients (NCT 021169882). Subjects were randomly assigned to receive 900 mg, 1350 mg, or 450 mg of oral rifampicin for 30 days. The study outcomes were PK data, safety/tolerability and efficacy (clinical response). In that study, we also collected information to investigate the inflammatory response, the neuroradiology response, and the role of gene-expert for TBM. The study is ongoing. We hope that the findings of that study will help us to understand TBM better and to provide valid evidence for the management of TBM.

5 FUTURE (POSSIBLE) RESEARCH

Finding the optimal dose of rifampicin is only one of the several strategic approaches in optimizing the treatment of TBM. Several questions can be addressed with respect to the available TB drugs. INH has a good penetration into the CSF (Donald, 2010); what is the PK profile of the current dose of INH in TBM patients (in plasma and in the CSF)? Is a higher dose of INH also needed in subjects with fast acetylator status? The same can be applied for pyrazinamide; this drug passes freely into the CSF (Donald, 2010). Data show that a higher dose of pyrazinamide might give a better outcome, but what is the risk of hepatotoxicity with higher dose of pyrazinamide in TBM patient?

Adjunctive therapy that involves the use of herbal medicine for TBM is one of the potential areas to be explored. The mechanism could be by enhancing the delivery of drugs to the CSF, inhibiting drug transporters (efflux pump) that play a role in low drug concentration in the CSF (Begley, 2004). Finding new drug(s) that have the ability to kill M. tb with an excellent ability to penetrate the blood–brain barrier and that are proved to be safe is the ultimate destination in drug discovery for TBM patients.

6 CONCLUSION

TBM is a real and devastating clinical problem. This disease kills patients and leave survivors with sequelae. Finding a better treatment of TBM is urgently needed to achieve a better treatment outcome. By understanding the pharmacokinetics of TB drugs, we show the evidence that the current dose of rifampicin, the backbone of TB treatment, is simply too low. A higher dose of rifampicin improved rifampicin exposure and tolerated by TBM patients, and it might improve treatment outcome. More research is needed to explore other innovative ways of optimizing TBM treatment. In doing so, it is important to build and maintain good inter-departmental and inter-faculty collaboration, both locally and internationally. Research roadmap is also very much important to keep us on track and to bring us to the eventual endeavor, i.e. a better patient care.

REFERENCES

Begley, D. J. 2004. Delivery Of Therapeutic Agents To The central nervous system: the problems and the possibilities. *Pharmacol Ther,* 104, 29–45.

Boeree, M. J., Diacon, A. H., Dawson, R., Narunsky, K., Du Bois, J., Venter, A., Phillips, P. P., Gillespie, S. H., Mchugh, T. D., Hoelscher, M., Heinrich, N., Rehal, S., Van Soolingen, D., Van Ingen, J., Magis-Escurra, C., Burger, D., Plemper Van Balen, G., Aarnoutse, R. E. & Pan, A. C. 2015. A dose-ranging trial to optimize the dose of rifampin in the treatment of tuberculosis. *Am J Respir Crit Care Med,* 191, 1058–65.

Donald, P. R. 2010. Cerebrospinal fluid concentrations of antituberculosis agents in adults and children. *Tuberculosis (Edinb),* 90, 279–92.

Ganiem, A. R., Parwati, I., Wisaksana, R., Van Der Zanden, A., Van De Beek, D., Sturm, P., Van Der Ven, A., Alisjahbana, B., Brouwer, A. M., Kurniani, N., De Gans, J. & Van Crevel, R. 2009. The effect of HIV infection on adult meningitis in Indonesia: a prospective cohort study. *AIDS,* 23, 2309–16.

Kanellakopoulou, K., Pagoulatou, A., Stroumpoulis, K., Vafiadou, M., Kranidioti, H., Giamarellou, H. & Giamarellos-Bourboulis, E. J. 2008. Pharmacokinetics of moxifloxacin in non-inflamed cerebrospinal fluid of humans: implication for a bactericidal effect. *J Antimicrob Chemother,* 61, 1328–31.

Ruslami, R., Ganiem, A. R., Dian, S., Apriani, L., Achmad, T. H., Van Der Ven, A. J., Borm, G., Aarnoutse, R. E. & Van Crevel, R. 2013. Intensified regimen containing rifampicin and moxifloxacin for tuberculous meningitis: an open-label, randomised controlled phase 2 trial. *Lancet Infect Dis,* 13, 27–35.

Savic, R. M., Ruslami, R., Hibma, J. E., Hesseling, A., Ramachandran, G., Ganiem, A. R., Swaminathan, S., Mcilleron, H., Gupta, A., Thakur, K., Van Crevel, R., Aarnoutse, R. & Dooley, K. E. 2015. Pediatric tuberculous meningitis: Model-based approach to determining optimal doses of the anti-tuberculosis drugs rifampin and levofloxacin for children. *Clin Pharmacol Ther,* 98, 622–9.

Shandil, R. K., Jayaram, R., Kaur, P., Gaonkar, S., Suresh, B. L., Mahesh, B. N., Jayashree, R., Nandi, V., Bharath, S. & Balasubramanian, V. 2007. Moxifloxacin, ofloxacin, sparfloxacin, and ciprofloxacin against Mycobacterium tuberculosis: evaluation of in vitro and pharmacodynamic indices that best predict in vivo efficacy. *Antimicrob Agents Chemother,* 51, 576–82.

Te Brake, L., Dian, S., Ganiem, A. R., Ruesen, C., Burger, D., Donders, R., Ruslami, R., Van Crevel, R. & Aarnoutse, R. 2015. Pharmacokinetic/pharmacodynamic analysis of an intensified regimen containing rifampicin and moxifloxacin for tuberculous meningitis. *Int J Antimicrob Agents,* 45, 496–503.

Yunivita, V., Dian, S., Ganiem, A. R., Hayati, E., Hanggono Achmad, T., Purnama Dewi, A., Teulen, M., Meijerhof-Jager, P., Van Crevel, R., Aarnoutse, R. & Ruslami, R. 2016. Pharmacokinetics and safety/tolerability of higher oral and intravenous doses of rifampicin in adult tuberculous meningitis patients. *Int J Antimicrob Agents,* 48, 415–21.

Iron-chelating effect of *Caesalpinia sappan* extract under conditions of iron overload

R. Safitri & D. Malini
Department of Biology, Faculty of Mathematics and Natural Sciences, Universitas Padjadjaran, Jl. Raya Bandung—Sumedang KM 21, Jatinangor, Sumedang West Java, Indonesia

A.M. Maskoen
Faculty of Dentistry, Universitas Padjadjaran, Jl. Raya Bandung—Sumedang KM 21, Jatinangor, Sumedang West Java, Indonesia

L. Reniarti, M.R.A.A. Syamsunarno & R. Panigoro
Faculty of Medicine, Universitas Padjadjaran, Jl. Raya Bandung—Sumedang KM 21, Jatinangor, Sumedang West Java, Indonesia

ABSTRACT: This study aims to obtain the effective dose of sappan wood extract (*Caesalpinia sappan* L.) that serves as a herbal chelating agent. An experiment with a Completely Randomized Design (CRD) was conducted on 21 male rats of 8 weeks old. The rats were given oral iron dextran and sappan wood extract at different doses for 15 days. Iron-related blood parameters were measured. The result revealed that a sappan wood extract dose of 200 mg/kg body weight had a chelating effect, showing a decline in ferritin levels (55.6%), a reduction in serum iron levels by 60%, and a reduction in transferrin saturation levels (84.7%). We also found an increase in transferrin levels (66.2%), and TIBC levels (62%) compared with rats given iron dextran injection alone. In conclusion, our study showed that a sappan wood extract dose of 200 mg/kg body weight has an ability to chelate excess iron in rats under conditions of iron overload.

1 INTRODUCTION

Thalassemia is a hereditary disease characterized by disturbances in the chain synthesis of hemoglobin or globin chains (Guyatt et al., 1990). The red blood cells in patients with thalassemia are short-lived, only about 1–2 months, so that the production of red blood cells is unbalanced and causes anemia due to impaired production of hemoglobin. In order to overcome this thalassemia, patients receive blood transfusions continuously. There is no process to remove excess iron in the body, which can be toxic (Guyatt et al., 1992).

An iron chelator is needed in patients with thalassemia to deal with the excess iron in the body. Iron chelator is a chelating agent that can bind to excess iron and then excrete them from the body (Aleem et al., 2014). Iron chelators that are commonly used by patients with thalassemia are Deferoxamine (Desferal), Deferiprone, and Deferasirox. Deferiprone and Deferasirox are active iron-binding agents when given orally. However, Deferoxamine medication is not convenient for patients with thalassemia, as it is given by subcutaneous infusion (below skin).

Deferoxamine medication often causes pain in patients with thalassemia, especially children, and hence, the compliance is low and it is quite difficult to apply. The other oral chelators such as Deferiprone and Deferasirox are rated impractical, uncomfortable, and are very expensive.

Sappan wood (*Caesalpinia sappan* L; Secang) is a plant that has high flavonoid content, which may be potential as a natural iron chelator of herbal origin. Organic components such as flavonoids may function as a metal chelator for their one-carboxyl group and an adjacent phenolic group reacts with metal ions to form a stable complex (Aleem et al., 2014).

In this study, the sappan wood extract was given at doses of 100, 200, and 400 mg/kg body weight to see the chelation ability of iron compared with another chelator, Deferiprone at a dose of 75 mg/kg body weight. The aim of this study is to investigate the potential role of sappan wood extract as an iron-chelating agent in rats (*Rattus norvegicus* L.) with excess iron condition. We proposed sappan wood extract as an iron chelator for patients with thalassemia.

2 MATERIALS AND METHODS

2.1 Instrument and materials

The equipment used in the study are: Atomic Absorption Spectrophotometry (AAS) variant types AA240FS, surgical equipment, hemocytometer, hemoglobinometer, oral syringes, microtiter plate reader (Biomerius), micropipette, Sartorius balance scales, pipette multi-channels, pipette capillary coated with heparin, rotary evaporator, centrifugator (Heraeus), and a spectrophotometer (Spectronic Genesys 8).

Materials used in the study are: distilled water, Deferiprone (Ferriprox), 96% ethanol, serum Fe Kit (Randox, Cat. No. SI 257), concentrated HCl, iron dextran, sappan wood, chloroform, feed pellet-type mice CP-551, PGA (Powder Arab GoM), rat Ferritin Kit (Abnova, Cat. No. KA 0211), rat Transferrin Kit (Abnova, Cat. No: KA 0510), TIBC Kit (Randox, Cat. TI 1010), and 21 male Wistar rats aged 8 weeks with an average weight of 200 g as test animals.

2.2 Preparation of sappan wood extract

The pieces of sappan woods were dried in the open air sheltered from direct sunlight and then pulverized to obtain crude drug sawdust. A wooden cup of powder weighing as much as 2.5 kg was fed into a Buchner funnel and then macerated using ethanol for 24 hours and repeated thrice. The macerate was filtered, and concentrated using a rotary evaporator at 40°C to obtain a thick extract. The viscous extract was then dried using a freeze dryer to obtain an extract in the form of crystalline powder.

2.3 Treatment provision

This study was approved by the Animal Care and Experimentation Committee in Faculty of Medicine, Padjadjaran University. Male rats that were 8 weeks old and weighing 200 g on average were used. Before commencing treatment, rats were acclimatized for a week. Acclimatization was done so that rats accustomed to the laboratory environment. During the acclimatization, rats were fed with the type of CP-551 laboratory chow and allowed to drink ad libitum. Rats were maintained in 12-hour light:12-hour darkness cycle and to keep the cage clean, reimbursement chaff was done 2 times a week.

Test compounds were administered orally using a gavage needle. Any substance that is given in advance is made in the form of a solution. Iron Dextran and Deferiprone first dissolved in distilled water, while sappan wood extract dissolved in the PGA. Treatment was given for 15 days, and iron dextran administered during an interval of 3 days. During the experiment, food and drink were available ad libitum.

The combination of treatment for animal testing is shown as follows (n = 3/group):

1. P0. Rats were given only distilled water (negative control)
2. P1. Rats were given only PGA (Powder Gum Arab)
3. P2. Rats were given only iron dextran at a dose of 60 mg/kg body weight (positive control)
4. P3. Rats were given only iron dextran at a dose of 60 mg/kg/ body weight and Deferiprone at a dose of 75 mg/kg/ body weight
5. P4. Rats were given only iron dextran at a dose of 60 mg/kg body weight and sappan wood extract at a dose of 100 mg/kg body weight
6. P5. Rats were given iron dextran at a dose of 60 mg/kg body weight and sappan wood extract at a dose of 200 mg/kg body weight
7. P6. Rats were given iron dextran at a dose of 60 mg/kg body weight and sappan wood extract at a dose of 400 mg/kg body weight

On day 16, all the test animals were killed using ketamine lethal dose, and blood samples were collected to measure ferritin, transferrin, iron levels in the liver, Total Iron Binding Capacity (TIBC), and iron levels in the serum. Blood was collected aseptically through the blood vessels in the neck. Blood sample (1.5 ml) was centrifuged at 1500 rpm for 15 minutes. The serum was taken and placed in a new microtube. Serums were stored at −20°C until used. As for the measurement of liver iron concentration (Hepatic Iron Concentration/HIC), the liver was isolated from rats aseptically and weighed freshly. Furthermore, it was inserted into the bottle containing physiological NaCl and stored at −20°C until use.

Ferritin levels were measured using Ferritin Kit (Catalog No. KA 0211), using the ELISA method. Transferrin levels were measured using Transferrin Kit (Catalog No. KA 0510), using the ELISA method. Examination of the liver iron content was conducted by AAS.

The level of TIBC and serum iron levels were examined using a spectrophotometer at a wavelength of 595 nm, using a kit from Randox (Laboratories, Cat. No. SI 257). However, before being tested with the kit, for the measurement of TIBC sample to be tested must be prepared in advance using a kit from Randox Laboratories, Cat. TI 1010. The measurement of serum levels of iron (Fe serum) was performed using a kit from Randox Laboratories, Cat. No. SI 257.

3 RESULTS AND DISCUSSION

3.1 Serum ferritin levels

Administration of iron dextran at a dose of 60 mg/kg body weight produced the highest ferritin levels at 3.92 ng/ml (Figure 1). Iron dextran increased ferritin levels by 65.6% compared with the control (distilled water treatment), suggesting iron overload condition. High ferritin levels indicate a high amount of iron stores in the body. Increased ferritin levels may occur due to liver cells damaged so much out of ferritin into plasma. This condition is in line with a previous study, that low ferritin levels indicate iron deficiency states, whereas ferritin levels have been found in circumstances such as hemochromatosis, thalassemia (repeated transfusions), and liver disease (Sheftel et al., 2009).

Rats with iron overload and Deferiprone at a dose of 75 mg/kg body weight showed an effect on lowering ferritin levels in rats under iron overload condition compared with a group with administration of iron dextran alone. This suggests that Deferiprone can effectively chelate iron. Deferiprone enters the cell and binds to the iron, which is then brought into the plasma. Deferiprone is an iron chelator bidentate (DFP takes three molecules to bind one iron atom), are lipophilic, uncharged, and has low molecular weight (MW = 212). Besides binding free iron in plasma, Deferiprone can easily penetrate cell membranes and bind iron in various intracellular organs compared with Deferoxamine (Hoffbrand et al., 2012).

Generally, it is known that sappan wood extract can reduce ferritin levels, which means that sappan wood extract can chelate iron. It can be seen through the sappan wood extract at 100, 200, and 400 mg/kg body weight dosage produces similar ferritin levels. Furthermore, the provision of PGA produces ferritin levels equal to sappan wood extract and distilled water intake. Giving sappan wood extract at 100 mg/kg body weight dose is effective as an iron chelator, as seen from the decline in serum ferritin. However, a decline in serum ferritin levels of rats that were given sappan wood extract at 200 mg/kg body weight dose is 55.6% lower than the administration of iron dextran so that the levels of ferritin are at 1.74 ng/ml. Low ferritin levels in sappan wood extract at 200 mg/kg body weight dose indicates that the extract can effectively have the ability to chelate iron that is better than Deferiprone.

Sappan wood extract has capabilities as an iron chelator allegedly because it has a high flavonoid content. Some flavonoids are efficient in chelating metal ion. Metals such as iron ion-free and copper-free can increase the formation of Reactive Oxygen Species (ROS) as seen in the formation of OH radicals in the following reaction: $H_2O_2 + Fe^{2+} (Cu^{2+}) \rightarrow OH\bullet + OH- + Fe^{3+} (Cu^{2+})$. Metal ion binding sites in flavonoids are the catechol group existing in ring B (Cao et al., 1997).

3.2 Transferrin levels

There was no difference between the levels of transferrin within test animals that were given distilled water with PGA (Figure 2). Rats that were given distilled water have transferrin levels of 4.44 ng/ml equal to rats given Deferiprone. In Deferiprone intake, there was an increase reaching 40.40% of the iron dextran. Provision of iron dextran to rats results in lowering of transferrin levels than the control given distilled water which is 3.82 ng/ml. The percentage decline reached 14%. This indicates that the rats in this study had sufficient iron level resulting in decreased levels of transferrin.

Free iron in plasma will catalyze the formation of free radicals. when they are not bounding with transferrin. Transferrin binds to receptors on the membrane of intestinal epithelial cells, and then

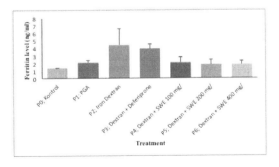

Figure 1. Ferritin levels (ng/dl) in rats that were given sappan wood extract under iron overload condition.

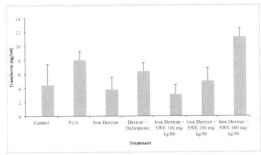

Figure 2. Transferrin levels (ng/ml) in rats with excess iron condition that were given sappan wood extract.

absorbed into the epithelial cells and released into the blood plasma in the form of plasma transferrin. Transferrin which had been pulled off the iron will come out through the receptor that will further capture the new iron, and so on (Dunn et al., 2003).

Rats that were given Deferiprone have transferrin levels close to the levels of transferrin in controls that were given distilled water at 6.41 ng/ml. Transferrin levels increase by 40.4% from the administration of iron dextran. High transferrin levels indicate that the iron is circulating in the plasma in very small amounts. This indicates that the administration of Deferiprone in this study is effective as an iron chelator in mice with the increased levels of transferrin.

Sappan wood extract at 100 mg/kg body weight dose produce transferrin levels as low as 3.14 ng/ml. This indicates that the extract sappan wood at 100 mg/kg body weight dose did not effectively bind iron; it was characterized by low levels of transferrin same as transferrin levels in the administration of iron dextran. Furthermore, the level of transferrin in rats that were given sappan wood extract at the dose of 400 mg/kg body weight is 66.2% higher than the administration of iron dextran at 11.29 ng/ml. Thus, the sappan wood extract at the dose of 400 mg/kg body weight in rats has the ability to chelate iron under iron overload conditions that can be seen from the rat transferrin levels, showing a reduction in high free iron in plasma.

3.3 *Hepatic iron levels*

Figure 3 shows that the administration of iron dextran, Deferiprone, and sappan wood extract altogether at 100, 200, and 400 mg/kg body weight doses produces high levels of iron in the liver. Rats injected with iron dextran at 60 mg/kg body weight dose had the highest iron content (102.3 ppm) and rats injected with distilled water had the lowest iron level (24.51 ppm). Injection of iron dextran can increase the iron content of the liver by 76.04%. This is supported by Bacon (1998) that liver is the main storage area of the body iron reserves (Bacon, 1998). The ability of iron to be involved in redox reactions can cause toxicity. This can occur if the body's iron storage capacity is exceeded.

In this study, Deferiprone showed the same ability to sappan wood extract administration at variation doses in lowering levels of hepatic iron in rats, which indicates that the Deferiprone chelates iron effectively in rats, both in plasma and tissue. Deferiprone has a function of iron chelation because it can penetrate the cell membrane, act as intracellular iron chelator, and can bind to free iron in plasma. In patients with thalassemia, Deferiprone is the first option because it is more practical, comfortable, and cheaper (Hoffbrand et al., 2012).

The intake of sappan wood extract at 100, 200, as well as 400 mg/kg body weight doses resulted in liver iron levels that are the same as the rats given iron dextran. Sappan wood extract at dose 400 mg/kg body weight reduced liver iron content by 24.01%. The higher dose of the sappan wood extract that was given to rats can reduce liver iron concentration indicating that SWE can effectively have the ability to chelate iron. Sappan wood extract contains flavonoids, which are the largest group of phenolic compounds. Shahidi (1995) mentioned that the flavonoids contained in sappan wood extract have capabilities to reduce or prevent the formation of hydroxyl free radical, superoxide anion, peroxyl radicals, and hydrogen peroxide (Shahidi et al., 2008). Selvaraj et al (2014) added that in addition to having antioxidant activity, flavonoids also have an ability to bind metal (Selvaraj et al., 2014).

The results of serum iron levels in Table 1 indicate that neither the Deferiprone, nor the variation doses of sappan wood extract (100, 200 and 400 mg/kg body weight) produce the same levels of serum iron. However, the administration of iron dextran produces the highest serum iron amounted to 842.01 mg/dl, an increase of 24.6%, much higher than the levels of serum iron in controls (distilled water). Iron overload rats with Deferiprone administration reduced iron levels by 57% from iron dextran-injected rats only. Iron dextran can also increase levels of serum iron and transferrin saturation. This was confirmed by Del Vecchio (2005) who stated that in patients with beta thalassemia major result of the destruction of erythrocytes stored iron in the cells of the Reticuloendothelial (RE), so the longer the time, the more the number of iron stored and RE cells' ability to store iron reduced. Therefore, iron will be separated into plasma which is then transported by transferrin (Del Vecchio et al., 2005).

Measurement of Total Iron-Binding Capacity (TIBC) is one of the parameters that used to

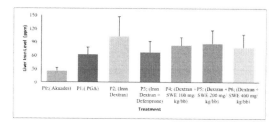

Figure 3. Liver iron levels (ppm) in rats that were given sappan wood extract under iron overload conditions.

Table 1. Mean of iron serum levels, Total Iron-Binding Capacity (TIBC), and transferrin saturation on rats with iron overload condition. The letters are different in one column indicating significant differences among treatments by the Duncan test (95%).

Treatments	Parameters		
	Serum Iron (µg/dl)	TIBC (µg/dl)	Saturation tf (%)
P0; Control	634,99[bc]	570,58[abc]	111,29[b]
P1; PGA	388,04[ab]	392,55[a]	98,85[ab]
P2; iron dextran	842,01[c]	262,21[a]	321,12[c]
P3; iron dextran + Deferiprone	362,4[a]	474,86[ab]	76,32[ab]
P4; Dextran + SWE 100 mg/kg body weight	338,8[a]	690,13[bc]	49,09[a]
P5; Dextran + SWE 200 mg/kg body weight	460,42[ab]	677,75[bc]	67,93[a]
P6; Dextran + SWE 400 mg/kg body weight	580,93[ab]	442,36[ab]	131,32[b]

see the capacity of iron bound in serum. When increase in serum iron happened under iron overload condition, TIBC levels will be declined. Table 1 shows that the administration of iron dextran, PGA, Deferiprone, distilled control, and sappan wood extract at 400 mg/kg body weight dose had the same levels of TIBC. Levels of TIBC in test animals decreased with the administration of iron dextran by 57.01% compared with the control.

TIBC is considered as a measure of Transferrin levels (Tf) in serum or plasma. TIBC (Total Iron-Binding Capacity) is the ability of transferrin to bind iron. Iron carried in the blood is attached to a protein called transferrin. Almost all of the iron in the serum binds to proteins, i.e., transferrin to TIBC indirectly showed levels of transferrin, which will increase the state of iron deficiency and decreased in the state of high serum iron or in a state of excess iron.

Transferrin saturation is calculated from the ratio between the concentrations of serum iron to TIBC expressed in percent (%). High transferrin saturation indicates high body iron reserves. Transferrin saturation values reached the highest in rats injected with iron dextran (312.12%). Iron overload rats with sappan wood extract at dose 100 mg/kg body weight dose had significant effects.

In Table 1, high iron level in serum, TIBC, and transferrin saturation due to iron dextran injection can be reduced by giving sappan wood extract from 100 to 400 mg/kg body weight dose. The optimum dose of sappan wood extract to reduce levels of serum iron and transferrin saturation is at 100 mg/kg body weight. It can be concluded that intake of sappan wood extract can reduce excess of iron even at low doses, suggesting the potential effect of Secang as an iron chelator.

4 CONCLUSIONS

Sappan wood extract (*Caesalpinia sappan* L.) can be used as an iron chelator with the ability to reduce ferritin by 55.6%, hepatic iron levels by 24.01%, serum iron levels by 60%, transferrin saturation by 84.7%, as well as increasing transferrin levels by 66.2% and TIBC by 62% in rats under conditions of iron overload. The dose of 200 mg/kg body weight has been found to be the effective dose of sappan wood extract as an iron chelator in rats under conditions of iron overload.

REFERENCES

Aleem, A., Shakoor, Z., Alsaleh, K., Algahtani, F., Iqbal, Z. & Al-Momen, A. 2014. Immunological evaluation of beta-thalassemia major patients receiving oral iron chelator deferasirox. *J Coll Physicians Surg Pak*, 24, 467–71.

Bacon, B. R. 1998. Iron overload states. *Clin Liver Dis*, 2, 63–75, vi.

Cao, G., Sofic, E. & Prior, R. L. 1997. Antioxidant and prooxidant behavior of flavonoids: structure-activity relationships. *Free Radic Biol Med*, 22, 749–60.

Del Vecchio, G. C., Nigro, A., Giordano, P. & De Mattia, D. 2005. Management of liver disease in thalassemia: main drug targets for a correct therapy. *Curr Drug Targets Immune Endocr Metabol Disord*, 5, 373–8.

Dunn, A., Carter, J. & Carter, H. 2003. Anemia at the end of life: prevalence, significance, and causes in patients receiving palliative care. *J Pain Symptom Manage*, 26, 1132–9.

Guyatt, G. H., Oxman, A. D., Ali, M., Willan, A., Mcilroy, W. & Patterson, C. 1992. Laboratory diagnosis of iron-deficiency anemia: an overview. *J Gen Intern Med*, 7, 145–53.

Guyatt, G. H., Patterson, C., Ali, M., Singer, J., Levine, M., Turpie, I. & Meyer, R. 1990. Diagnosis of iron-deficiency anemia in the elderly. *Am J Med,* 88, 205–9.

Hoffbrand, A. V., Taher, A. & Cappellini, M. D. 2012. How I treat transfusional iron overload. *Blood,* 120, 3657–69.

Piperno, A. 1998. Classification and diagnosis of iron overload. *Haematologica,* 83, 447–55.

Selvaraj, S., Krishnaswamy, S., Devashya, V., Sethuraman, S. & Krishnan, U. M. 2014. Flavonoid-metal ion complexes: a novel class of therapeutic agents. *Med Res Rev,* 34, 677–702.

Shahidi, F., Mcdonald, J., Chandrasekara, A. & Zhong, Y. 2008. Phytochemicals of foods, beverages and fruit vinegars: chemistry and health effects. *Asia Pac J Clin Nutr,* 17 Suppl 1, 380–2.

Sheftel, A. D., Richardson, D. R., Prchal, J. & Ponka, P. 2009. Mitochondrial iron metabolism and sideroblastic anemia. *Acta Haematol,* 122, 120–33.

The role of *S.aureus* and *L.plantarum* as an immunomodulator of IFNα macrophages and fibronectin dermal fibroblast secretion

R.S.P. Saktiadi
Postgraduate Program, Faculty of Medicine, Universitas Padjadjaran, Bandung, Indonesia
Department of Biology, Indonesia University of Education, Bandung, Indonesia

S. Sudigdoadi, T.H. Madjid & E. Sutedja
Postgraduate Program, Faculty of Medicine, Universitas Padjadjaran, Bandung, Indonesia

R.D. Juansah
Department of Biology, Indonesia University of Education, Bandung, Indonesia

T.P. Wikayani & N. Qomarilla
Cell Culture Laboratory, Faculty of Medicine Universitas Padjadjaran, Bandung, Indonesia

T.Y. Siswanti
Molecular Genetic Laboratory, Faculty of Medicine Universitas Padjadjaran, Bandung, Indonesia

ABSTRACT: The aim of the research is to determine the levels of IFNα macrophages and fibronectin fibroblast after adding *S. aureus* and *L. plantarum*. The research used rat peritoneal macrophages and fibroblast dermal cell cultures. Heat killed *S. aureus* and *L. plantarum* was made a suspension at dosages of 10^6, 10^7, 10^8 cells/mL. The macrophages was exposed to SA for 6 hours, continued with SALP treatment, and LP for 24 hours, and then supernatant was separated for treatment in fibroblast for 24 hours before measured by ELISA. The data was analyzed using ANOVA and Duncan MRT at a CI of 95%. In general, SA exposure increased IFNα and fibronectin (p < 0,000) greater than LP, while LP exposure decreased fibronectin in proportion to the increases in dosage. It could be concluded that high dose *S.aureus* increased fibronectin greater than *L. plantarum*. Both *S. aureus* and *L. plantarum* increased IFNα. Conversely, exposures of *S.aureus* with *L. plantarum* simultaneously decreased it.

Keywords: immunomodulator, IFNα, fibronectin, *L. plantarum*, *S.aureus*

1 INTRODUCTION

Atopic Dermatitis (AD) is a chronic skin disease characterized by barrier dysfunction and skin inflammation. The skin of AD patients is highly prone to bacterial colonization, and *Staphylococcus Aureus*-(SA) is the most frequently found bacterium. *S.aureus* infection apparently plays a significant role in the pathogenesis of disease, both as the cause and as exacerbation of skin inflammation. The advantage of combined antibiotic-corticosteroid therapy against *S.aureus* is often temporary in nature and resistance may come to be a serious problem in long-term therapy. Therefore, further microbiological analyses are needed to find an adequate, effective approach. (Cork *et al.* 2009., Pascolini *et al.* 2011., Birnie *et al.* 2008., Cogen *et al.* 2008).

A current research on *S.aureus* strain expressed a different structure of surface protein, known as Microbial Surface Components Recognizing Adhesive Matrix Molecules (MSCRAMMs), which recognize Extracellular Matrix (ECM) proteins in human. It was known that the MSCRAMMs family is bonded specifically to ECM proteins such as Fibronectin and collagen, thus involving the protein as a potential ligand for the bonding of *S.aureus* to the skin of an individual with DA.(Cho *et al.* 2001a., Cho *et al.* 2001b., Faunholz *et al.* 2012).

Fibronectin is a glycoprotein of a high molecular weight that is present in an extracellular matrix, bonded to a membrane protein receptor called integrin. In addition to integrin, fibronection is bonded to numerous other host molecules, as well as non-host molecules, such as fibronectin binding proteins bacterial-derived (e.g., FBP-A, FBP-B, domain N-terminal). (Williams *et al.* 2002., Collona. 2007).

IL4, a proinflammation cytokine, also increases the fibronectin deposit of suprabasal epidermis area. It affirms a selective mechanism where Th2 response can increase S.aureus bond to skin by inducing fibronectin production, providing an inflammation environment to atopy and mediating the increase of S.aureus attachment to skin. (Collona. 2007., Haileselassie et al. 2013).

A research found that probiotics may reduce inflammation and ameliorate atopic dermatitis. Manipulation of the amount and activation of inflammatory cells can be made as an approach to controlling during inflammatory and healing processes. Various cytokines that immune cells produce exert a deep effect on fibroblast migration, proliferation, and production of matrix extracellular. Earlier, an in vitro research had found that an interaction between inflammatory cells and fibroblast can modulate some fibroblast functions, including collagen and fibronectin productions. (Williams et al. 2002., Lebeer et al. 2012., Prince et al. 2012).

Probiotic mechanism of *Lactobacilli* and *S.aureus* on IFNα and Fibronectin secretion in pathogenesis and atopic dermatitis healing process is not fully understood. Therefore, the present research was aimed to determine the levels of macrophage-produced IFNα and that of fibroblast cell-produced fibronectin after adding *L.plantarum* and *S.aureus*.

2 MATERIALS AND METHODS

2.1 *S. aureus* and *L. plantarum* cultures

An ATCC25923-derived *S.aureus* was cultured in a Mueller Hinton and incubated at 37°C for 24 hours in an aerobic condition. An ATCC8014-derived *L.plantarum* was cultured in a MRS medium and incubated at 37°C for 24 hours in an anaerobic condition. The cultures were harvested at a *mid-log* phase and then a bacterial suspension was made by comparing it against McFarland standard, adjusted till reached $10^6, 10^7, 10^8$ cfu/mL (*S-aureus*-SA) and $10^6, 10^7, 10^8$ cfu/mL (*L.plantarum*-LP), and confirmed by spectrophotometer. Furthermore, a killing process was done by a heat-killed process, by heating in a waterbath at 70°C for 30 minutes, and then left at room temperature and stored at 4°C until its use.

2.2 Peritoneal macrophages culture

Macrophage cells obtained from *Wistar* rats, were washed 3 times, counted, and their viability was assessed with Trypan blue. Afterward, the macrophages was cultured in RPMI + 3% FCS in an incubator 5% CO_2 at 37°C for 45 minutes to allow macrophages attaching to the plate; then macrophages was displaced into a number of wells, 0.5×10^6 cells/well. The treatment may begin 4–24 hours after plating.

2.3 Stimulation of macrophage cells with S.aureus and treatment of L.plantarum

The overnight incubated peritoneal macrophages cell culture was then treated by SA at dosages of $10^6, 10^7, 10^8$ cells/mL for 6 hours. Supernatant was removed and washing was done by RPMI twice, and then LP was added at dosages of $10^6, 10^7, 10^8$, cells/mL, incubated for 24 hours. Then, the supernatant was transferred into an ELISA plate for the measurement of the macrophages cytokine levels.

2.4 Skin dermal fibroblast culture

A sample of Wistar rat skin that has been minced 2–3 mm² in size was transferred into a plate that contained DMEM + 10% FCS; the fibroblast cells attached were cultured in an incubator 5% CO_2; at 37°C till achieved confluence. The third passage fibroblasts that have assesed the viability of which has been examined was then transferred in a number of wells, 0.5×10^5 cells/well, and now ready to do treatment with the supernatant of macrophages culture for 24 hour. Then, the supernatant was separated for measuring its fibronectin levels.

3 RESULTS

Based on the research result as presented on Fig. 1, the highest treatment dosages of both *S.aureus* and *L.plantarum* exposures (SA3-1a and LP3-1c, respectively) increased IFNα levels, where the most significant were SA3 by 51.40 pg/dL and LP3 by 33.97 pg/dL ($p < 0.000$). However, the exposure of high-dosage combined *S.aureus* and *L.plantarum* (SALP3-1b) decreased significantly the levels of IFNα macrophages (17.50 pg/dL).

Whereas an increase in fibronectin levels due to higher dosage SA (1d) causes an increase in the levels of fibronectin (from 82.50 ng/dL to 92.84 ng/dL), an exposure of LP (1f) to skin dermal fibroblast cell culture by a high dosage causes the decreasing levels of fibronectin fibroblast (from 86.98 ng/dL to 76.08 ng/dL) ($p < 0.000$). Meanwhile, the exposure of combined SALP (1e) caused a lower increase of the levels of fibronectin fibroblast compared by LP and SA exposure respectively.

4 DISCUSSION

In this study, increasing dosages of SA exposure to macrophages cell culture caused the increase of IFNα in proportion to the increase of dosage. The highest levels of interferon was yielded by an exposure of high dosage SA, 51.40 pg/dL. This espoused the findings of some similar researches on

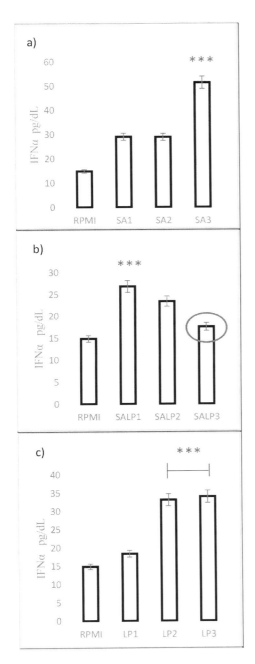
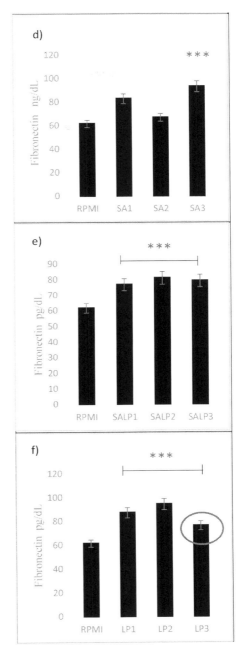

Figure 1. (*Continued*).

the modulation of immune system due to *S. aureus* colonization, by CpG molecules to be recognized particularly by TLR2, which then induces the productions of proinflammation cytokines such as TNFα, IL6 by macrophages through MyD88-dependent signaling pathways that stimulates IRF7

Figure 1. IFNα macropages levels (a, b, and c), and Fibronectin fibroblasts levels (c, d, and f); activation with *S. aureus*-SA (SA and SALP) and without activation with SA (LP). 24 h supernatant harvested from macrophages and fibroblasts cultures, IFNα macropages and Fibronectin fibroblasts measured in supernatants from triplicate cultures in one represantative of six experiments. Bars represent means and SEM ***p < 0.000 different from medium control.

(IFN-regulatory factor-7) and finally induces IFNα pathways.(Pascolini et al. 2011).

Moreover, an increase in LP dosage also increased IFNα levels by 33.97 pg/dL, probably due to an increase of antigen presentation in form of CpG molecules through TLR4 and TLR9 on macrophages cell surface, which then increased the releases of proinflammation cytokines and inductions of type-I IFN secretion like IFNα, that might prevent infection in health cells. (Pascolini et al. 2011., Cogen et al. 2008., Cho et al. 2001).

Innate immunity system represented by macrophage cells enhanced defense against pathogenic infection in the increasing levels of IFNα that served to stimulate p53 activity and hence cell apoptosis, playing a role in preventing infection of health cells. Other function of interferon is to increase the regulation of MHC I and MHC II molecules, in addition to increasing *immunoproteasome* activity. The high expression of MHC I increases the presentation of antigen to cytotoxic T cells, while *immunoproteasome* processes the loading of antigen to MHC I molecules, thus enhancing the recognition of antigen and killing infected cells. Meanwhile, the high expression of MHC II also increases the presentation of antigen to helper T cells, in turn increasing the releases of various cytokines such as various interferons and interleukins that promote signaling and coordinate other immune cells. The increase of IFNα levels by the exposure of *L.plantarum* is less than that of IFNα levels by the exposure of *S.aureus*.

It is apparently disadvantageous given that IFNα is very needed in the apoptotic process of microbe-infected cells. In other side, however, because *L.plantarum* is not phagocyted by cells, the presence of a high IFNα levels is not a requirement. (Cogen et al. 2008., Cho et al. 2001a., Cho et al. 2001b).

Conversely, in the exposure of combined SALP, high dosage caused a decreasing content of IFNα to be 17.50 pg/dL. This may resulted from excessive presentation of antigen due to the protein originated from both microbes (CpG), which activating different signaling pathways, eventually increasing the contents of anti-inflammatory cytokines like TGFβ and IL10 by macrophages that caused decreases in the releases of inflammatory cytokines including IFNα. It indicated that *S.aureus* exposure simultaneously with *L.plantarum* to macrophage cells may be beneficial. Therefore, the administration of *L.plantarum* as a probiotic therapy in overcoming skin inflammation cases such as in AD patients may be investigated further. (Cho et al. 2001a., Cho et al. 2001b., Faunholz et al. 2012).

Fibronectin. The highest increase of fibronectin levels due to SA exposure was at the highest dosage, SA3. It was probably because *S.aureus* colonization increased the presentation of antigen by macrophage cells which in turn promoted proinflammation cytokines like IL4 and anti-inflammatory cytokine like TGFβ due to high dosages of pathogens (10^8 cells/dL). Both IL4 and TGFβ have some extensive function to fibroblast cells, i.e., to promote fibronectin synthesis. (Williams et al. 2001).

Unlike SA, LP exposure to macrophages cell culture decreased the levels of fibronectin yielded by fibroblast in proportion to the increases in dosage (LP3 caused the lowest fibronectin synthesis). It was likely due to increases in either TGFβ antiinflammation or IL10, thus causing a down-regulation leading to decreases in immune activity and hence releases of proinflammatory cytokines. Inflammatory process began decreasing and hence the production of fibronectin by fibroblast cell culture. In addition, decrease in fibronectin levels might also be caused by the far lower pathogenity nature of *L.plantarum* than that of *S.aureus*. The pathogenity of a microbe antigen are caused by several things, among others 'strangeness' of an antigen. The more strange an antigen, the higher the immune response is activated, hence very likely the higher the cytokine released. The strangeness of an antigen molecule can be recognized by among others unmethylated CpG DNA which is often present in the DNA of bacteria. Genus *Lactobacilli* falls into a category of low CpG, meaning low content of CpG molecule. That makes possible for the lower immune system activation of *L.plantarum* than that of *S.aureus*. The low capacity of immune cell (macrophages) activation causes equally low production and secretion of proinflammatory cytokine, and hence low secretion of fibronectin levels. High levels of fibronectin is actually disadvantageous for both the course of AD disease and its healing process, because a high fibronectin, especially for long time, deteriorates inflammatory process resulting from fibronectin as a potential ligand on *S.aureus* surface antigen. (Colonna. 2007., Haileselassie et al. 2013., Lebeer et al. 2012., Prince et al. 2012).

An exposure of combined SALP increases the fibronectin levels of fibroblast. However, compared to the increase of fibronectin due to SA and LP respectively, the increase of fibronectin by SALP is the lowest. It seems that there is a mechanism that makes it possible for obstructing the activation of immune system by either anti-inflammatory cytokines or a new signaling pathways that may suppress the production of fibronectin due to the exposure of the two microbes. The earlier researches by some experts have found that *Lactobacilli* may stimulate the secretion of anti-inflammatory cytokine such as TGFβ and IL10, eventually decreasing the production of fibronectin by fibroblast and fibronectin deposit in skin epidermis. Based on our research, it seemingly possible for further research on an application of probiotic therapy particularly

L.plantarum in dealing with AD patients. (Cho et al. 2001b., Lebeer et al. 2012).

S.aureus; *L.plantarum*; Fibronectin, IFNα, and atopic dermatitis. The high levels of fibronectin due to SA exposure in our research led to increasing colonization of SA on the skin of AD patients, deteriorating the condition of inflammation that already existed in AD. This espoused the findings of earlier researches concerning the effect of SA infection on the progressiveness of AD disease. High fibronectin levels also increases the induction of autophagy by the nonprofessional cells of phagocyte such as fibroblast, keratinocyte, epithelium, and endothelium that may cause the persistence of *S.aureus* in cells and is able to avoid apoptosis. (Faunholz et al. 2012).

On the other side, the exposure of LP in the present research reduced fibronectin synthesis by the culture of fibroblast cells, and increased IFNα synthesis by the culture of macrophage cells. This could prevent the persistent infection of intracellular *S.aureus* by the high levels of IFNα capable of increasing the apoptotic process of infected cells, and prevent increase of fibronectin caused the obstruction of autophagy by cells, and thus there was no intracellular infection, and also prevent persistence of bacteria on skin. (Soong et al. 2015).

Other interesting finding was that the increase of fibronectin production by fibroblast resulting from exposure of combined SA and LP was lower than that of SA and LP respectevely. It indicated the role of LP in inhibiting the increase of fibronectin production. The decreasing fibronectin synthesis by the exposure of LP in the present research came to be a new hope for further *in vivo* research relating to the application of LP to shorten the phase of inflammation on patients with atopic dermatitis.

5 CONCLUSION

Based on our research, it could be concluded that the exposure of *L.plantarum* caused a lower increase in the levels of fibronectin fibroblast than the exposure of *S.aureus*. Both *L.plantarum* and *S.aureus* respectively, increased the synthesis of IFNα macrophages, while the exposure of *S.aureus* together with *L.plantarum* decreased the levels of IFNα macrophages. Therefore, the use of *L.plantarum* as a probiotic therapy in cases of skin inflammation such as atopic dermatitis, could be further investigated.

REFERENCES

Birnie AJ, Bath-Hextall FJ, Ravenscroft JC, Williams HC. Interventions to reduce Staphylococcus aureus in the management of atopic eczema. Cochrane Database of Systematic Reviews 2008, Issue 3. Art. No.: CD003871. DOI: http://doi.org/10.1002/14651858.CD003871.pub2.

Cho, S. H., Strickland, I., Tomkinson, A., Fehringer, A. P., Gelfand, E. W., & Leung, D. Y. M. (2001). Preferential binding of Staphylococcus aureus to skin sites of Th2-mediated inflammation in a murine model. *Journal of Investigative. Dermatology*, 116(5), 658–663. http://doi.org/10.1046/j.0022-202X.2001.01331.x.

Cho, S. H., Strickland, I., Boguniewicz, M., & Leung, D. Y. M. (2001). Fibronectin and fibrinogen contribute to the enhanced binding of Staphylococcus aureus to atopic skin. *Journal of Allergy and Clinical Immunology*, 108(2), 269–274. http://doi.org/10.1067/mai.2001.117455.

Cogen, A. L., Nizet, V., & Gallo, R. L. (2008). Skin microbiota: A source of disease or defence? *British Journal of Dermatology*, 158(3), 442–455. http://doi.org/10.1111/j.1365-2133.2008.08437.x.

Colonna, M. (2007). TLR pathways and IFN-regulatory factors: To each its own. *European Journal of Immunology*, 37(2), 306–309. http://doi.org/10.1002/eji.200637009.

Cork, M. J., Danby, S. G., Vasilopoulos, Y., Hadgraft, J., Lane, M. E., Moustafa, M., Ward, S. J. (2009). Epidermal barrier dysfunction in atopic dermatitis. *The Journal of Investigative Dermatology*, 129(8), 1892–1908. http://doi.org/10.1038/jid.2009.133.

Faunholz, M., & Sinha, B. (2012). Intracellular Staphylococcus aureus: live-in and let die. *Frontiers in Cellular and Infection Microbiology*, 2(43), 1–7. http://doi.org/10.3389/fcimb.2012.00043.

Haileselassie, Y., Johansson, M. A., Zimmer, C. L., Björkander, S., Petursdottir, D. H., Dicksved, J., Sverremark-Ekström, E. (2013). Lactobacilli Regulate Staphylococcus aureus 161:2-Induced Pro-Inflammatory T Cell. Responses. In.Vitro. *PloS. ONE*, 8(10), 1–12. http://doi.org/10.1371/journal.pone.0077893.

Lebeer, S., Claes, I. J. J., & Vanderleyden, J. (2012). Anti inflammatory potential of probiotics: Lipoteichoic acid makes a difference. *Trends in Microbiology*, 20(1), 5–10. http://doi.org/10.1016/j.tim.2011.09.004.

Pascolini, C., Sinagra, J., Pecetta, S., Bordignon, V., De Santis, A., Cilli, L., Ensoli, F. (2011). Molecular and immunological characterization of Staphylococcus aureus in pediatric atopic dermatitis: Implications for prophylaxis and clinical management. *Clinical and Developmental Immunology*, 2011. http://doi.org/10.1155/2011/718708.

Prince, T., McBain, A. J., & O'Neill, C. A. (2012). Lactobacillus reuteri protects epidermal keratinocytes from Staphylococcus aureus-induced cell death by competitive exclusion. *Applied and Environmental Microbiology*, 78(15), 5119–5126. http://doi.org/10.1128/AEM.00595-12.

Soong G, Paulino F, Watchel S, Parker D, Wickersham M, Zhang D, et al. 2015. Methicilin-resistant S. aureus adaptation to human keratinocytes. mbio.asm.org. 1(2):1–15.

Williams, R. J., Henderson, B., Sharp, L. J., & Nair, S. P. (2002). Identification of a Fibronectin-Binding Protein from Staphylococcus epidermidis Identification of a Fibronectin-Binding Protein from Staphylococcus epidermidis. *Infect Immun*, 70(12), 6805–10. http://doi.org/10.1128/IAI.70.12.6805.

Advances in Biomolecular Medicine – Hofstra, Koibuchi & Fucharoen (Eds)
© 2017 Taylor & Francis Group, London, ISBN 978-1-138-63177-9

Exon globin mutation of β-thalassemia in Indonesian ethnic groups: A bioinformatics approach

N.I. Sumantri, D. Setiawan & A. Sazali
Master Program of Biotechnology, Universitas Padjadjaran, Bandung, Indonesia

ABSTRACT: HbVar (http://globin.bx.psu.edu/hbvar/) and HOPE (http://www.cmbi.ru.nl/hope/) are relational databases for genetic disorders, especially hemoglobinopathies. Genetic variants of thalassemia have impacts on its severity. Here, we report exon β-globin mutations in Indonesian ethnic groups, including structural and functional analysis. HbVar identified three codons, namely Cd 15 (G>A), Cd 17 (A>T), and Cd 30 (AGG>AGC), with transition and transversion mutations. Structural and functional alterations were analyzed using HOPE. Mutants mainly result in hydrophobicity alteration. Functionally, heme, iron, and oxygen binding of globin depend on the type of mutation. Some databases provide information about thalassemia and its related aspects. Updates of inter-related databases depend on later mutation found in populations, which will be useful for research, diagnostic, and therapeutic purposes.

Keywords: HbVar, HOPE, β-thalassemia, Indonesian ethnic groups

1 INTRODUCTION

Thalassemia is one of the major hemoglobinopathies among the population all around the world. It is considered as the most widespread genetic mutation (Ablahad, A., 2013). According to the World Health Organization (WHO), 1.5 to 7% of the world population are carriers of this disease. Moreover, about 60,000–400,000 new cases have been reported each year. This mutation prevents the production of any or some β-globin. The lack of β-globin impels to minimize the amount of operative hemoglobin and excessive crushing of red blood cells, which can lead to anemia (Sarnaik, S.A., 2005). Thalassemia has become a major public health issue in Indonesia. It has been estimated that up to 10% of the population carries a gene associated with β-thalassemia (Widyanti, et al., 2011). The number of individuals suffering from this severe, life-shortening disorder is increasing. Thalassemia is common in Indonesia, where 13 mutations of β-thalassemia have been identified, of which the most prevalent are the HbE, IVS-nt5, and Cd 35 mutations (Setianingsih, et al., 1998). In this paper, we focus on the transition and transversion mutations of the exon β-globin in Indonesian ethnic groups.

HbVar (http://globin.bx.psu.edu/hbvar) is one of the oldest and most appreciated locus-specific databases launched in 2001 by a multi-center academic effort to provide timely information on genomic alterations leading to hemoglobin variants and all types of thalassemia and hemoglobinopathies (Giardine, B., et al., 2014). Phenotypic descriptions, associated pathology, and ethnic occurrences, accompanied by mutation frequencies and references, are included in this database. Results obtained from HbVar can be analyzed by other databases, such as FINDbase (http://www.findbase.org), Leiden Open-Access Variation database (http://www.lovd.nl), PhenCode (http://phencode.bx.psu.edu/), and HOPE (http://www.cmbi.ru.nl/hope/). HOPE is a free-access automatic program that analyzes the structural and functional effects of point mutations in the protein-coding region of the human genome. HOPE helps life scientists and biomedicists to understand how mutation affects human physiology by displaying readable 3D mutant structures (Vanselaar, et al., 2010). In this study, we report structural and functional alterations of β-globin mutations of β-thalassemia in Indonesian ethnic groups.

2 METHODS

Data of exon β-globin mutations in Indonesian ethnic groups were collected from HbVar (http://globin.bx.psu.edu/hbvar/, accessed in July, 29th 2016). A number of columns on query page were filled according to the objective (Fig. 1). Variants with transversion and transition mutations were selected to be analyzed for structural and functional alterations using HOPE (http://www.cmbi.ru.nl/hope/,

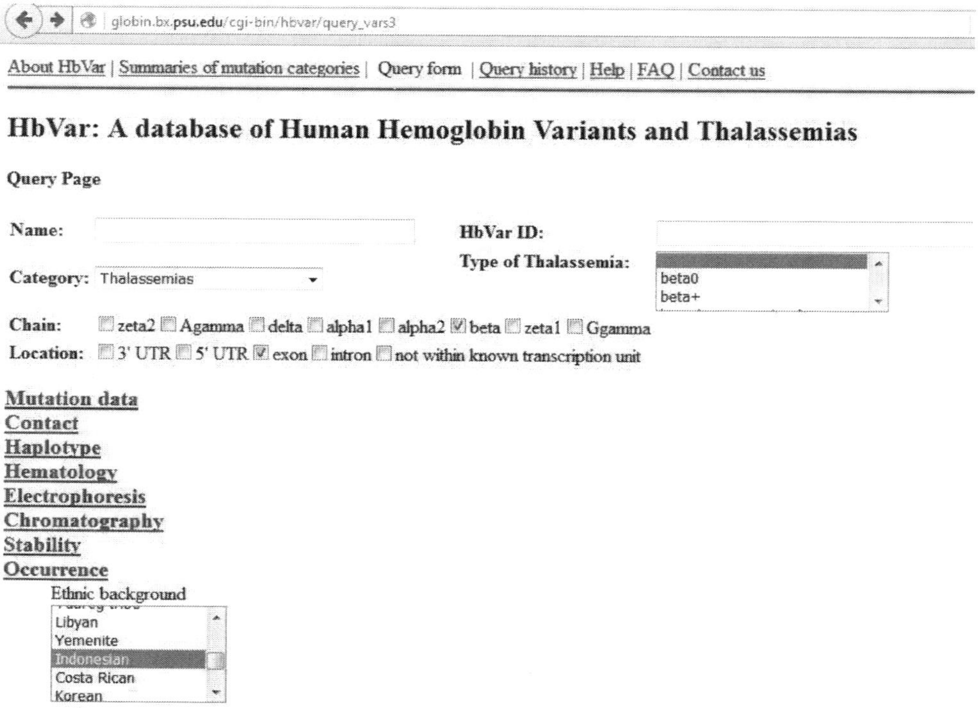

Figure 1. Query page in HbVar to find some information about thalassemia and other hemoglobinopathies.

accessed in July, 30th 2016). Results were described descriptively.

3 RESULTS AND DISCUSSION

3.1 *Exon globin mutation of β-thalassemia*

The point mutations causing β-thalassemia result from single-base substitutions, insertions or deletions of a few bases within introns or exons might affect the splicing pattern of the pre-mRNAs (Thein, 1998; Thein, 2004). According to the results of our study, there were 5 exon mutations of β-thalassemia in Indonesian ethnic groups, but only 3 codons were with transversion and transition mutations (Fig. 2). Cd 15 and Cd 30 were transition mutants, whereas Cd 17 was a transversion mutant. Those three mutations were β⁰-thalassemia.

Based on the HbVar database, Cd 15 was the highest occurrence (6.78%), followed by Cd 17 and Cd 30 (each 1.69%). Cd 15, Cd 17, and Cd 30 were common mutations of β-thalassemia in Asia (Yatim, 2014; Datkhille, 2015; Sultana, 2016). Lie-Enjo (1989) showed that there were a number of common β-thalassemia mutations in Indonesia, including HbE (Cd 26 GAG->AAG), IVS-1 nt5 (G->C), IVS-2 nt654 (C->T), Cd 41/42 (deletion CTTT), codon 17 (AAG->TAG), IVS-1 nt1 (G->T), IVS-1 nt1 (G->A), Cd 15 (TGG->TAG), Cd 30 (AGG->ACG), and Cd 35 (deletion C).

3.2 *Structural and functional alteration of mutant globin*

Structural and functional alterations were analyzed using HOPE. This database collects information from a wide range of information sources including calculations on the 3D-coordinates of the protein using WHAT IF Web services, sequence annotations from the UniProt database, conservation scores from HSSP, and predictions by a series of Distributed Annotation System (DAS) services. When possible, homology models are built with YASARA. Data are stored in a database and combined in a decision scheme to identify the effects of a mutation on the protein's 3D structure and its function. The report will be shown at the same website and is illustrated with figures and animations, showing the effects of the mutation (Camilli, F., et al., 2011). Globins are heme-containing proteins that are best known for their roles in oxygen (O_2) transport

Figure 2. β-Thalassemia with exon globin mutations in Indonesian ethnic groups.

and storage. Globin gene mutations affect hemoglobin (Hb), the major blood oxygen (O_2) carrier. These mutations are broadly subdivided into those that impair globin protein subunit production (thalassemia) and those that produce structurally abnormal globin proteins (Hb variants). Naturally occurring Hb mutations cause a range of biochemical abnormalities, some of which produce clinically significant symptoms (Thom, CS., et al., 2013). Some structurally abnormal β-globin-chain variants are also quantitatively reduced, with a phenotype of β-thalassemia, in which case they are referred to as "thalassemicemoglobinopathies". Deletions or insertions of entire codons allow the reading frame to remain in phase, and the remaining amino acids are normal. In all cases, no trace of abnormal hemoglobin could be detected by the standard techniques of IEF, HPLC, or heat stability tests. One mechanism that could explain the lack of detection of these structural β-globin-chain variants is that the affected amino acids are involved in α1 β1 contacts (Thein, SL., 2013). Definitive hemoglobin identification can be performed by protein sequencing, DNA analysis, and HPLC combined with electrospray mass spectrometry in a specialized reference laboratory (Mubeen, H., et al., 2016).

The mutant residue is smaller than the wild-type residue. The wild-type residue was positively charged, and the mutant residue is neutral. The mutant residue is more hydrophobic than the wild-type residue. The wild-type residue forms a hydrogen bond with glutamic acid at position 122. The difference in hydrophobicity will affect hydrogen bond formation. The wild-type residue forms a salt bridge with glutamic acid at position 122. The difference in charge will disturb the ionic interaction made by the original, wild-type residue. The mutated residue is located on the surface of a domain with unknown function. The residue was not found to be in contact with other domains, the function of which is known within the used structure. However, contact with other molecules or domains is still possible and might be affected by this mutation.

The mutant structure of CD 15 is shown in Figure 3.

The mutant residue is smaller than the wild-type residue. The mutation will cause an empty space in the core of the protein, mutating the wild-type residue into a residue type that is considered non-natural. The mutated residue is located in a domain that is important for binding of other molecules and in contact with residues in a domain that is also important for binding. The mutation might disturb the interaction between these two domains and as such affect the function of the protein. This domain is annotated with the following Gene-Ontology terms to indicate its function oxygen and heme binding.

Figure 4 shows the structure alteration of CD 17.

The wild-type residue was positively charged, and the mutant residue is neutral (Fig 5). The difference in hydrophobicity will affect hydrogen bond formation. The wild-type residue forms a salt bridge with glutamic acid at position 27. The difference in charge will disturb the ionic interaction made by the original, wild-type residue. The mutation introduces a smaller residue at this position. The new residue might be too small to make multimeric contacts. A more hydrophobic residue is introduced here. Any hydrogen bond that could be made by the wild-type residue to other monomers will be lost now and affect the multimeric contacts.

Codon 15 (G->A); TGG(Trp)->TAG(stop codon)

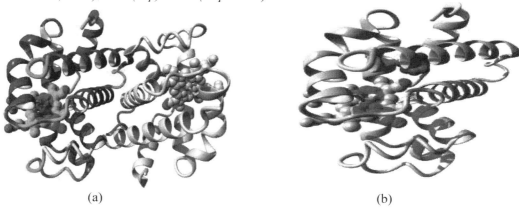

(a) (b)

Figure 3. Mutant globin codon 15 (G->A) in the 3D-structure. (a) Overview of protein in ribbon-presentation. The protein is indicated by colored according to elements: α-helix is blue, β-strand is red, turn is green, 3/10 helix is yellow, and random coil is cyan. Other molecules in the complex are colored grey when present. (b) Overview of protein in ribbon-presentation. The protein is colored grey, the side chain of the mutated residue is colored magenta and shown as small balls.

Codon 17 (A->T); AAG(Lys)->TAG(stop codon)

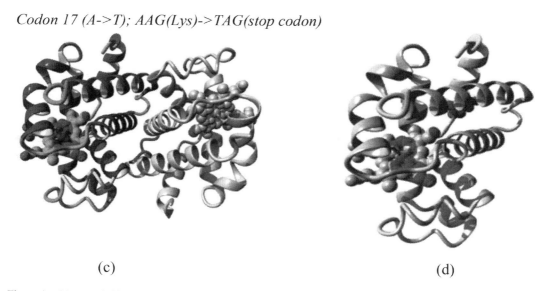

(c) (d)

Figure 4. Mutant globin codon 17 (A->T) in the 3D-structure. (c) Overview of the protein in ribbon-presentation. The protein is colored by element; α-helix = blue, β-strand = red, turn = green, 3/10 helix = yellow, and random coil = cyan. Other molecules in the complex are colored grey when present. (d) Overview of the protein in ribbon-presentation. The protein is colored grey, the side chain of the mutated residue is colored magenta and shown as small balls.

The mutated residue is not in direct contact with a ligand; however, the mutation could affect the local stability which in turn could affect the ligand contacts made by one of the neighboring residues. The wild-type residue forms a hydrogen bond with glutamic acid at position 27. This residue is part of an Interpro domain named hemoglobin β-type.

This is a strong indication that the residue is indeed in contact with other proteins. The mutated residue is located in a domain that is important for binding of other molecules and in contact with residues in a domain that is also important for binding. The mutation might disturb the interaction between these two domains and as such affect the function

Codon 30 (AGG->AGC) [IVS-I-130 (+1)]

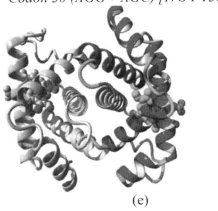
(e)

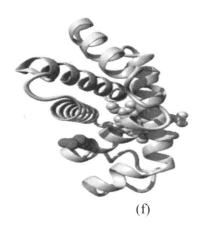
(f)

Figure 5. Mutant globin codon 30 (AGG->AGC) in the 3D-structure. (e) Overview of the protein in ribbon-presentation. The protein is colored by element; α-helix = blue, β-strand = red, turn = green, 3/10 helix = yellow, and random coil = cyan. Other molecules in the complex are colored grey when present. (f) Overview of the protein in ribbon-presentation. The protein is colored grey, the side chain of the mutated residue is colored magenta and shown as small balls.

of the protein. This domain is annotated with the following Gene-Ontology (GO) terms to indicate its function: heme, oxygen, and iron ion binding.

Preventive diagnosis and follow-up would reduce infant mortality by preventing the development of severe anemia as well as dangerous complications. Bioinformatics approaches show the mutations in the HBB gene. Protein sequence analysis shows various domains in selected sequence. Studying sequence variations and analyzing these at the molecular level can help overcome many genetic disorders.

4 CONCLUSION

We found three codons with transition and transversion mutations in β-thalassemia in Indonesian ethnic groups. These mutated residues result in different sizes from the wild type and mainly disturb protein activity by structural alteration. A complete spectrum of genetic lesions affecting the β-globin gene that gives rise to a spectrum of phenotypic severity has been described. Some databases provide information about thalassemia and its related aspects. Updates of inter-related databases depend on later mutation found in populations, which will be useful for research, diagnostic, and therapeutic purposes.

ACKNOWLEDGMENTS

The authors gratefully acknowledge the generosity of lecturers in the master program of the biotechnology department for their worthy support and inspirational discussion, without whom the present study could not have been completed.

REFERENCES

Camilli, F, Borrmann A, Gholizadeh S, Beek T te, Kuipers R, Venselaar H. The future of HOPE: what can and cannot be predicted about the molecular effects of a disease causing point mutation in a protein? *EMBnet. Journal.* 2011.17(1).

Datkhile, KD, Patil MN, Vavhal RD, and Khamkar TS. The Spectrum of b-Globin Gene Mutations in Thalassemia Patients of South-Western Maharashtra: A Cross Sectional Study. *JKIMSU.* 2015. 4. 3.

Giardine B, Borg J, Viennas E, Pavlidis C, Moradkhani K, Joly P, Bartsakoulia M, Riemer C,Miller W, Tzimas G, Wajcman H, Hardison RC, andPatrinos GP. Updates of the HbVar database of human hemoglobin variants and thalassemia mutations. *Nucleic Acids Research.* 2014. 42.

Lie-Injo LE, Cai SP, Wahidijat I, Moeslichan S, Lim ML, Evangelista L, Doherty M, and Kan YW. Beta-thalassemia mutations in Indonesia and their linkage to beta haplotypes. *Am J Hum Genet.* 1989 Dec; 45(6): 971–975.

Mubeen, H, Naqvi RZ, Masood A, Shoaib MW and Raza S. In silico mutation analysis of human beta globin gene in sickle cell disease patients. Int J Res Med Sci. 2016. 4(5): 1673–1677.

Sarnaik, S.A. Thalassemia and Related Hemoglobinopathies. Indian Journal of Pediatrics, Volume 72—April, 2005.

Setianingsih, I.I. Molecular basis of beta-thalassemia in Indonesia: application to prenatal diagnosis. Molecular Diagnosis, 3(1) 1998: 11–19.

Sultana, GNN, Begum R, Akhter H, Shamim Z, Rahim MA, and Chubey G. The Complete Spectrum of Beta (β) Thalassemia Mutations in Bangladeshi Population. *Austin Biomark Diagn* 3. 1.

Thein, S.L. Beta thalassaemia. In: Baillie`re's Clinical Haematology, Sickle Cell Disease and Thalassaemia (ed. by G.P. Rodgers), Baillie`re Tindall, London. 1998. 11. 1: 91–126.

Thein, SL. Genetic insights into the clinical diversity of ß-thalassaemia. *Br J Hematol*. 2004. 124: 264–274.

Thein, SL. The Molecular Basis of β-thalassemia. *Cold Spring Harb Perspect Med*. 2013. 3.

Thom, CS, Dickson CF, Gell DA, and Weiss MJ. Hemoglobin Variants:Biochemical Properties and Clinical Correlates. *Cold Spring Harb Perspect Med*. 2013. 3.

Venselaar, H, te Beek TAH, Kuipers RKP, Hekkelman ML, and Vriend G. Protein structure analysis of mutations causing inheritable diseases. An e-Science approach with life scientist friendly interfaces. *BMC Bioinformatics*. 2010. 11. 548.

Widayanti, C.G., Ediati, A., Tamam, M., Faradz, S.M.H., Sistermans, E.A., Plass, A.M.C. Feasibility of preconception screening for thalassaemia in Indonesia: exploring the opinion of Javanese mothers. Ethnicity & Health: 2011, 16(4–5), 483–499.

Yatim, NFM, Rahim MA, Menon K, Al-Hassan FM, Ahmad R, Manocha AB, Saleem M, and Yahaya BH. Molecular Characterization of α- and β-Thalassaemia among Malay Patients. *Int. J. Mol. Sci.* 2014. 15. 8835–8845.

Serum immunoglobulin-E level correlates with the severity of atopic dermatitis

O. Suwarsa, E. Avriyanti & H. Gunawan
Department of Dermatology and Venereology, Faculty of Medicine, Universitas Padjadjaran—Dr. Hasan Sadikin Hospital, Bandung, West Java, Indonesia

ABSTRACT: The Immunoglobulin E (IgE) plays an important role in Atopic Dermatitis (AD) pathogenesis, in principle, it could serve as a marker for AD. Though, in some studies, showed no evidence of their correlation. The severity of AD is commonly evaluated using the Scoring Atopic Dermatitis (SCORAD).This study aim was to assess the correlation between serum IgE levels and severity of AD. Serum from 20 AD outpatients was collected and their IgE levels were evaluated using Enzyme-Linked Immunosorbent Assay (ELISA) technique. The SCORAD assessments were also performed. There were significant differences serum IgE levels between AD and control group ($p < 0.05$). Furthermore, the correlation analysis result between serum IgE level and SCORAD was significant ($r = 0.73$; $p < 0.05$). The result showed that the increased of IgE level indicated the increasing of SCORAD.We suggest serum IgE level proved to be a useful tool in assessing the severity of AD.

Keywords: atopic dermatitis, immunoglobulin E, SCORAD

1 INTRODUCTION

Atopic Dermatitis (AD) is a chronic, relapsing, inflammatory skin disease characterized by eczematous skin lesions, with itchy sensations (Leung et al. 2008). This disease affects 20–30% of children and 5–10% of adults in industrialized countries.

Scores to assess the severity of many skin diseases are important, particularly for evaluation of treatments and classification purposes (Stalder & Taieb 1993). Scores of AD have been developed to determine disease severity through an evaluation of lesion characteristics such as erythema, induration, and scaling in relation to the affected body surface area(Bieber 2008, 2010). The most widely used assessmentfor AD severityis the Scoring Atopic Dermatitis (SCORAD) (Stalder & Taieb 1993).

The Immunoglobulin-E (IgE), IgE-mediated mast cell, and eosinophil activationcontribute to AD pathogenesis, but direct evidence supporting this mechanism is still unclear (Liu et al. 2011).Typically, AD patients tend to have greatly elevated levels of IgE (Bieber 2008) and IgE could be a marker for AD. However, some studies showed no evidence ofa correlation between IgE elevated levels and the degree of severity in AD (Gerdes 2009).

Thus, in this study, we tried to analyse the correlation between serum IgE levels and severity of AD.

2 MATERIALS AND METHODS

2.1 Patients

Twenty AD patients were enrolled using a consecutive sampling from Dermatology Clinic, Dr. Hasan Sadikin Hospital and fulfilled the Hanifin and Rajka criteria. The 20 healthy subjects were chosen as a control group. The subjects were excluded from the study when they used systemic and/or topical corticosteroids, antibiotics, and other immunosuppressive agents, such as cyclosporine, tacrolimus, azathioprine, etc. within the last two weeks before the study was started.They were also excluded if atopic manifestations exist, such as asthma, allergic rhinitis, or hay fever.

The study was approved by the Health Research Ethics Committee, Faculty of Medicine, Universitas Padjadjaran—Dr. Hasan Sadikin Hospital, Bandung, Indonesia. Oral explanation about the study was done and subsequently written informed consent was obtained from each participant.

2.2 Scoring Atopic Dermatitis (SCORAD)

An SCORAD analysis from the European Task Force on Atopic Dermatitis was used to determined the AD severity based on clinical symptoms (Stalder & Taieb 1993). Classification of the SCORAD scores are mild (<15), moderate (15–40), and severe (>40) (Liu et al. 2011).

2.3 Enzyme-Linked Immunosorbent Assay (ELISA) for IgE

The IgE level was measured from subject's peripheral blood by ELISA sandwich method using anti-human IgE monoclonal antibodies conjugated with peroxidase (Enzygnost-IgE monoclonal) (BioMérieux, France) according to the manufacturer's instructions. The levels of the IgE in IU/ml in each patient were used for data analysis.

2.4 Statistical analysis

Results were expressed as mean ± SD and statistically significant differences were defined as a *p value* less than 0.05. The Mann-Whitney analysis was used for determination of differences of IgE level in the AD and control group. The Pearson correlation test was applied for correlation analysis of IgE with SCORAD.

3 RESULTS

Figure 1 showed the mean level of serum IgE in AD patients was 2,666.31 ± 2230.51 IU/ml, and in the control group was 247.85 ± 364.86 IU/ml. There were statistically significant differences between serum Ig levels AD and control group (p < 0.05).

The mean serum IgE level in mild, moderate, and severe AD patient based on SCORAD were 536.83 ± 282.08 IU/ml, 1,975.19 ± 1,235.98 IU/ml, and 4,566.26 ± 2482.3 IU/ml, respectively (Fig. 2). Furthermore, the correlation analysis result between serum IgE level and SCORAD was r = 0.73 and p < 0.05. These results suggested that the increased of IgE level indicated the increasing of SCORAD (Fig. 3).

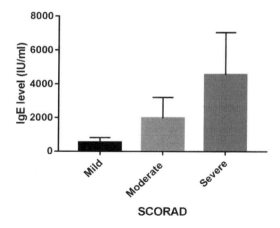

Figure 2. The level of IgE in AD patients with mild, moderate, and severe SCORAD. The data represent the averages ± SD.

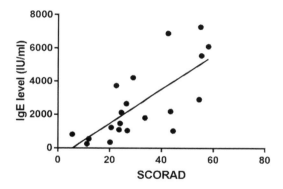

Figure 3. Correlation of IgE and severity of AD score showed a significant result (r = 0.73; p < 0.05).

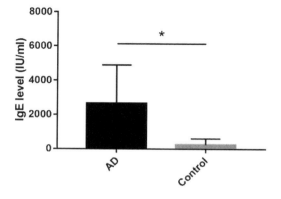

Figure 1. The level of IgE in AD and control group showed a significant difference using Mann-Whitney analysis. The data represent the averages ± SD. *p < 0.05.

4 DISCUSSION

The assessment of severity of AD validated scores has been established. For AD, the SCORAD is most commonly used. Besides these, many clinical instruments have been made to establish serum markers for disease severity and activity in AD (Bunikowski et al. 1999).

The activation of immune cells, especially T lymphocytes, B lymphocytes, eosinophil, and mast cells, and their migration to the skin have an essential role in the AD pathogenesis (Werfel et al. 2016). A study suggested that patients with AD showed increasing serum IgE levels (Laske & Niggeman 2004). Our results also indicate a correlation between the severity of AD and serum IgE levels. In our study, serum IgE level in AD patients was higher than the control group

(Fig. 1).These results were similar with Campbell et al. study, that total IgE levels were 4,434 IU/ml with a range of 60–54,296 IU/ml in 74 AD patients (Campbell & Kemp 1998). Novak et al. (2005) also showed that the average levels of total IgE in 225 AD patients was 1,399 KU/L. Other studies also reported that serum IgE of AD patients have a higher value relatively with the control (Aral et al. 2006, Brenninkemeijer et al. 2005, Orfali et al. 2009, Chernyshov 2009).

According to minor aspects of Hanifin-Rajka criteria, the increase in total serum IgE is a minor sign of AD, although it has no predictive value for long-term prognosis (Orfali et al. 2009). A complex interaction between genetic, environmental, skin barrier, pharmacologic, and immunologic factors, have contributed to the pathogenesis of AD. Many studies showed the increase of interleukin (IL)-4, IL-5 and IL-13 cytokines, while levels of IFN-γ decrease in the peripheral blood of AD patients (Novak & Bieber 2008). Hence, IL-4 and IL-13 induce the production of IgE by B cells (Kang et al. 2003, Novak & Bieber 2008). Most people with AD have increased eosinophilia and serum IgE levels (Liu et al. 2011).

In the previous study, total serum IgE levels were reported to be associated with AD and severity of the disease, thus being a possible marker of AD activity (Dhar et al. 2005, Jones et al. 1975). Increased IgE serum levels were found in individuals with more severe AD (Iju 1985, Laske & Niggemann 2004). Therefore, it fits well that we observed a higher and statistically highly significant correlation between IgE and SCORAD (Fig. 3).

5 CONCLUSION

In conclusion, our results suggest that serum IgE levels correlate with the degree of severity in AD and serum IgE level might be a useful tool in assessing the severity of AD.

REFERENCES

Aral, M., Arican, O., Gul, M., Sasmaz, S., Kocturk, S.A., Kastal, U., Ekerbicer, H.C. 2006. The relationship between serum levels of total IgE, IL-18, IL-12, IFN-gamma and disease severity in children with atopic dermatitis. *Mediators of Inflammation* 4: 73–98.

Bieber, T. 2008. Atopic dermatitis. *The New England Journal of Medicine* 358:1483–1494.

Bieber, T. 2010. Atopic dermatitis. *Annals of Dermatology* 22:125–137.

Brenninkemeijer, E., Spuls, P., Legierse, C., Lindeboom, R., Smitt, H., Bos, J. 2008. Clinical differences between atopic and atopiform dermatitis. *Journal of the American Academy of Dermatology* 58:407–414.

Bunikowski, R., Mielke, M., Skarabis, H., Herz, U., Bergmann, R.L., et al. 1999. Prevalence and role of serum IgE antibodies to the *Staphylococcus aureus*-derived superantigen SEA and SEB in children with atopic dermatitis. *Journal of Allergy and Clinical Immunology* 103:119–124.

Campbell, D.E. & Kemp, A.S. 1998. Production of antibodies to staphylococcal superantigens in atopic dermatitis. *Archive of Disease in Childhood* 79:400–404.

Chernyshov, P.V. 2009. B7–2/CD28 costimulatory pathway in children with atopic dermatitis and its connection with immunoglobulin E, intracellular interleukin-4 and interferon-gamma production by T cells during a 1-month follow-up. *Journal of the European Academy of Dermatology and Venereology* 23:656–659.

Dhar, S., Malakar, R., Chattopadhyay, S., et al. 2005. Correlation of the severity of atopic dermatitis with absolute eosinophil counts in peripheral blood and serum IgE levels. *Indian Journal of Dermatology, Venereology and Leprology* 71:246–249.

Gerdes S, Kurrat W, Mrowietz U. 2009. Serum mast cell tryptase is not a useful marker for disease severity in psoriasis or atopic dermatitis. *British Journal of Dermatology* 160:736–40.

Iju, M. 1985. Study of the genesis of atopic dermatitis-atopic dermatitis and IgE. *Hokkaido Igaku Zasshi* 60: 806–833.

Jones, H.E., Inouye, J.C., McGerity, J.L., et al. 1975. Atopic disease and serum immunoglobulin-E. *British Journal of Dermatology* 92:17–25.

Kang, K., Polstr, A.M., Nedorost, S.T., Steven, S.R., Cooper KD. 2003. Atopic dermatitis. In: Bolognia, J.L., Rapini, R.P. (eds), *Dermatology*: 199–214. Philadelphia: Elsevier.

Laske, N. & Niggemann, B. 2004. Does the severity of atopic dermatitis correlate with serum IgE levels?. *Pediatric Allergy and Immunology* 15:86–88.

Leung, D.Y.M., Eichenfield, L.F., Boguniewics, M. Atopic dermatitis (atopic eczema). 2008. In: Wolff, K., Goldsmith, L.A., Katz, S.I., Gilchrest, B.A., Paller, A.S., Leffel, D.J. (eds), *Fitzpatrick's dermatology in general medicine*: 146–157. 7th ed. New York: McGraw-Hill.

Liu, F.T., Goodarzi, H., Chen, H.Y. 2011. IgE, mast cells, and eosinophils in atopic dermatitis. *Clinical Reviews in Allergy and Immunology* 41:298–310.

Novak, N. & Bieber, T. 2008. The pathogenesis of atopic dermatitis. In: Reitamo, S., Luger, T.A., Steinhoff, M., (eds). *Textbook of atopic dermatitis*: 25–31. 1st ed. London: Informa Healthcare.

Novak, N., Kruse, S., Potreck, J., Maintz, L., Jenneck, C., Weidenger, S., Fimmers, R., Bieber, T. 2005. Single nucleotide polymorphisms of the IL-18 gene are associated with atopic eczema. *Journal of Allergy and Clinical Immunology* 115:828–833.

Oranje, A.P., Glazenburg, E.J., Wolkerstorfer, A., de Waard-van der Spek, F.B. 2007. Practical issues on interpretation of scoring atopic dermatitis: the SCORAD index, objective SCORAD and the three-item severity score. *British Journal of Dermatology* 157:645–648.

Orfali, R.L., Sato, M.N., Takaoka, R., Azor, M.H., Rivitti, E.A., Hanifin, J.M., Aoki, V. 2009. Atopic dermatitis in adults: evaluation of peripheral blood mononuclear cells proliferation response to *Staphylo-

coccus aureus enterotoxins A and B and analysis of interleukin-18 secretion. *Experimental Dermatology* 18:628–633.

Stalder, J.F. & Taieb, A. 1993. European Task Force on Atopic Dermatitis. Severity scoring of atopic dermatitis: the SCORAD Index. *Dermatology* 186:23–31.

Werfel T, Allam JP, Biedermann T, Eyerich K, Gilles S, Guttman-Yassky E, et al. 2016. Cellular and molecular immunologic mechanisms in patients with atopic dermatitis. *Journal of Allergy and Clinical Immunology* 138:336–349.

Fatty liver in fasted FABP4/5 null mice is not followed by liver function deterioration

M.R.A.A. Syamsunarno
Department of Biochemistry and Molecular Biology, Faculty of Medicine Universitas Padjadjaran, West Java, Indonesia
Department of Medicine and Biological Science, Gunma University Graduate School of Medicine, Maebashi, Gunma, Japan

M. Ghozali, G.I. Nugraha & R. Panigoro
Department of Biochemistry and Molecular Biology, Faculty of Medicine Universitas Padjadjaran, West Java, Indonesia

T. Iso
Department of Medicine and Biological Science, Gunma University Graduate School of Medicine, Maebashi, Gunma, Japan

M. Putri
Department of Medicine and Biological Science, Gunma University Graduate School of Medicine, Maebashi, Gunma, Japan
Department of Biochemistry, Universitas Islam Bandung, Bandung, Indonesia

M. Kurabayashi
Department of Medicine and Biological Science, Gunma University Graduate School of Medicine, Maebashi, Gunma, Japan

ABSTRACT: We have recently found that deletion of Fatty Acid-Binding Protein-4 and -5 (FABP4/5) resulted in a marked perturbation of metabolism in response to fasting, including fatty liver. The purpose of our study was to investigate liver function in FABP4/5 null mice (DKO mice) with fatty liver after prolonged fasting. Wild-Type (WT) and DKO mice were fed and fasted 24 hours and 48 hours. Liver was collected and preserved for Hematoxylin and Eosin (HE) and Masson's Trichrome (MT) staining. To analyze liver function, serum was collected to determine Aspartate Transaminase (AST) and Alanine Transaminase (ALT) levels. After 48 hours of fasting, there was no increase in fibrosis in the liver of DKO mice. The serum level of AST was lower in DKO mice while that of ALT was comparable after 48 hours of fasting, suggesting that liver damage in DKO mice was modest. In conclusion, the study shows that massive lipid accumulation in fasted DKO mice is not accompanied by liver function deterioration.

Keywords: FABP4, FABP5, fasting, liver function, fatty liver

1 INTRODUCTION

The emerging problem of fatty liver has become a major concern in research. The prevalence of fatty liver is 10 to 24% in general population and the number increases from 57 to 74% in obese people (Angulo, 2002). Fatty liver is associated with metabolic disorder, including hypertriglyceridemia and insulin resistance. The pathophysiology of fatty liver into inflammatory stage, Non-Alcoholic Steatohepatitis (NASH), has been proposed in "two hits" hypothesis (Tessari et al., 2009). The first hit is insulin resistance leading to fatty liver disease, followed by the second hit, the condition which involved oxidative stress, determining lipid peroxidation, increased cytokine production, and inflammation leading to NASH. It is well known that the susceptibility of advanced fibrosis or cirrhosis leads to death in patients with fatty liver (Tessari et al., 2009, George and Liddle, 2008).

Fasting can promote triglyceride accumulation in the liver, suggested as an adaptive mechanism in response to the lack of energy source (Teusink et al., 2003). The liver is a central organ of energy metabolism to control the energy balance. In the fasting condition, when the intake of carbohydrates is diminished, the first line of energy source is glucose derived from liver glycogen. Then, fatty acid is released from the adipose tissue and utilized by the liver or peripheral organs, such as heart and skeletal muscle. FA should be transported to the liver for beta-oxidation or partially oxidized to produce ketone bodies.

The protein carrier facilitates fatty acid transport from the vascular lumen through the endothelium into the extracellular matrix tissue. Fatty Acid-Binding Proteins (FABPs) family are small fatty acid-binding molecules that exert an influence on the physiology function after interacting with the transcription factor gene or the regulator gene (See review in (Furuhashi et al., 2011)). Among those FABPs families, FABP4 (adipocyte-FABP/aP2) is distributed in adipocyte and macrophage, while FABP5 (Keratinocyte-FABP/Mal1) is widely distributed in various cells including keratinocyte and macrophage. Both FABPs show similar expression in some cells, suggesting that they have an interaction under particular circumstances. Indeed, when FABP4 function is diminished, FABP5 expression is highly increased as a compensation function. Therefore, FABP4/5 Double-Knockout (DKO) model is necessary to gain deep understanding of the function of these FABPs. Fasting FABP4 knockout mouse showed triglyceride accumulated in the liver (Baar et al., 2005).

In our previous study, we have showed that FABP4/5 was expressed in muscular capillary endothelial type of cells and serve to facilitate fatty acid transport from vascular lumen into the extracellular matrix (Iso et al., 2013). Deletion of FABP4/5 resulted in a marked perturbation of metabolism in response to prolonged fasting, including hyperketotic hypoglycemia, hepatic steatosis, and inability to maintain body temperature under cold exposure (Syamsunarno et al., 2013, Syamsunarno et al., 2014). The purpose of the current study is to investigate whether massive triglyceride accumulation in fasted DKO mice could induce liver function disturbance.

2 MATERIALS AND METHODS

2.1 *Mice and sample collection*

Mice deficient for both *Fabp4* and *Fabp5* (DKO mice) were generated from an intercross between *Fabp4*(-/-) and *Fabp5*(-/-) mice, as described previously (Maeda et al., 2005). The background strain of the wild-type and knockout mice was C57BL6J.

The Institutional Animal Care and Use Committee (Gunma University Graduate School of Medicine) approved all study protocols. The mice were housed in a temperature-controlled room in a 12-hour light/12-hour dark cycle and had unrestricted access to water and standard chow (CE-2, Clea Japan, Inc.). Mice from both genotypes that were 12 to 18 weeks old were used. Fasting experiments were conducted as described previously (Syamsunarno et al., 2013). Mice were individually housed and food was withdrawn for 24 or 48 hr; water was provided ad libitum. The liver was processed for histological examinations and blood was collected from the inferior vena cava and centrifuged at 1,500xg for 15 minutes at 4°C to separate the serum.

2.2 *Liver histology*

Liver samples were fixed with 4% paraformaldehyde and embedded in paraffin. The liver was stained with Hematoxylin and Eosin (HE) and Masson's trichrome, as described elsewhere (Sunaga et al., 2013).

2.3 *Aspartate aminotransferase and alanine aminotransferase test*

Serum levels of Aspartate Aminotransferase (AST) and Alanine Aminotransferase (ALT) were enzymatically measured by using L-Type Wako AST-J2 and ALT-J2, respectively (Wako Chemical, Osaka).

2.4 *Statistical analysis*

Statistical analysis was performed using ANOVA for three samples or more and Bonferroni's *post hoc* multiple comparison tests were performed to evaluate differences between the control and experimental groups. Significant difference was determined as $P<0.05$. Data are presented as means ± S.E.

3 RESULTS AND DISCUSSION

Lipid accumulation in the liver might reflect inflammation and potentially deteriorates liver function. However, neither inflammatory cell infiltration nor increase in fibrosis was observed in the liver of DKO mice after prolonged fasting (Figures 1 and 2). The serum level of AST was lower in DKO mice while that of ALT was comparable after 48 hours of fasting (Figures 3 and 4). These results suggested that liver damage in DKO mice was modest.

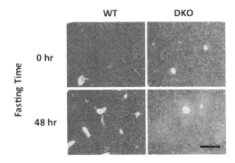

Figure 1. Representative HE staining of the liver of WT and DKO mice. Scale bar = 100 μm.

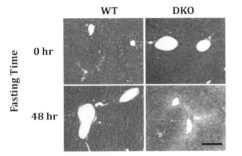

Figure 2. Representative Masson's trichrome staining of the liver of WT and DKO mice. Scale bar = 100 μm.

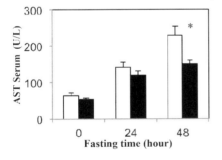

Figure 3. Serum levels of AST before and after fasting (0 or 48 hours). n = 8/group. *p < 0.05.

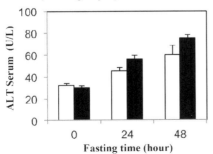

Figure 4. Serum levels of ALT before and after fasting (0 or 48 hours). n = 8/group. *p < 0.05.

4 DISCUSSION

The liver has a capacity to store the excess of fatty acids in triglyceride form. Then, these fatty acids will be oxidized or exported back into the main vascular system in the Low-Density Lipoprotein (LDL) form. There are two main functions of fatty acid oxidation. First, to generate energy that will be used to sustain hepatocyte function and gluconeogenesis. Second, fatty acid will be partially oxidized to produce ketone bodies, where 3-hydroxy-3-methylglutaryl-CoA synthase 2 (HMGCS2) is a key regulator gene.

In our previous study, the accumulation of triglyceride in the liver of DKO mice after prolonged fasting was higher compared with the liver of WT mice (Syamsunarno et al., 2013). Additionally, fatty liver condition in DKO mice is reversible after re-fed with normal chow (Syamsunarno et al., 2013). In the current study, fibrosis sign or increased AST and ALT are not detected. These findings indicate that remarkable lipid accumulation in the liver of DKO mice is not accompanied by inflammation and severe destruction of hepatocytes.

5 CONCLUSIONS

This study shows that massive lipid accumulation in the liver of FABP4/5 DKO mice after prolonged fasting is not accompanied by liver function deterioration.

REFERENCES

Angulo, P. 2002. Nonalcoholic fatty liver disease. *N Engl J Med,* 346, 1221–31.

Baar, R. A., Dingfelder, C. S., Smith, L. A., Bernlohr, D. A., Wu, C., Lange, A. J. & Parks, E. J. 2005. Investigation of in vivo fatty acid metabolism in AFABP/aP2(-/-) mice. *Am J Physiol Endocrinol Metab,* 288, E187–93.

Furuhashi, M., Ishimura, S., Ota, H. & Miura, T. 2011. Lipid chaperones and metabolic inflammation. *Int J Inflam,* 2011, 642612.

George, J. & Liddle, C. 2008. Nonalcoholic fatty liver disease: pathogenesis and potential for nuclear receptors as therapeutic targets. *Mol Pharm,* 5, 49–59.

Iso, T., Maeda, K., Hanaoka, H., Suga, T., Goto, K., Syamsunarno, M. R., Hishiki, T., Nagahata, Y., Matsui, H., Arai, M., Yamaguchi, A., Abumrad, N. A., Sano, M., Suematsu, M., Endo, K., Hotamisligil, G. S. & Kurabayashi, M. 2013. Capillary endothelial fatty acid binding proteins 4 and 5 play a critical role in fatty acid uptake in heart and skeletal muscle. *Arterioscler Thromb Vasc Biol,* 33, 2549–57.

Maeda, K., Cao, H., Kono, K., Gorgun, C. Z., Furuhashi, M., Uysal, K. T., Cao, Q., Atsumi, G., Malone, H., Krishnan, B., Minokoshi, Y., Kahn, B. B., Parker, R. A. & Hotamisligil, G. S. 2005. Adipocyte/macrophage fatty acid binding proteins control integrated metabolic responses in obesity and diabetes. *Cell Metab,* 1, 107–19.

Sunaga, H., Matsui, H., Ueno, M., Maeno, T., Iso, T., Syamsunarno, M. R., Anjo, S., Matsuzaka, T., Shimano, H., Yokoyama, T. & Kurabayashi, M. 2013. Deranged fatty acid composition causes pulmonary fibrosis in Elovl6-deficient mice. *Nat Commun,* 4, 2563.

Syamsunarno, M. R., Iso, T., Hanaoka, H., Yamaguchi, A., Obokata, M., Koitabashi, N., Goto, K., Hishiki, T., Nagahata, Y., Matsui, H., Sano, M., Kobayashi, M., Kikuchi, O., Sasaki, T., Maeda, K., Murakami, M., Kitamura, T., Suematsu, M., Tsushima, Y., Endo, K., Hotamisligil, G. S. & Kurabayashi, M. 2013. A critical role of fatty acid binding protein 4 and 5 (FABP4/5) in the systemic response to fasting. *PLoS One,* 8, e79386.

Syamsunarno, M. R., Iso, T., Yamaguchi, A., Hanaoka, H., Putri, M., Obokata, M., Sunaga, H., Koitabashi, N., Matsui, H., Maeda, K., Endo, K., Tsushima, Y., Yokoyama, T. & Kurabayashi, M. 2014. Fatty acid binding protein 4 and 5 play a crucial role in thermogenesis under the conditions of fasting and cold stress. *PLoS One,* 9, e90825.

Tessari, P., Coracina, A., Cosma, A. & Tiengo, A. 2009. Hepatic Lipid Metabolism And Non-Alcoholic Fatty liver disease. *Nutr Metab Cardiovasc Dis,* 19, 291–302.

Teusink, B., Voshol, P. J., Dahlmans, V. E., Rensen, P. C., Pijl, H., Romijn, J. A. & Havekes, L. M. 2003. Contribution of fatty acids released from lipolysis of plasma triglycerides to total plasma fatty acid flux and tissue-specific fatty acid uptake. *Diabetes,* 52, 614–20.

The potential of seluang fish (*Rasbora* spp.) to prevent stunting: The effect on the bone growth of *Rattus norvegicus*

Triawanti
Medical Chemistry/Biochemistry Department, Faculty of Medicine, Lambung Mangkurat University, Banjarmasin, Indonesia

A. Yunanto
Pediatric Department, Faculty of Medicine, Lambung Mangkurat University, Banjarmasin, Indonesia
Ulin General Hospital, Banjarmasin, Indonesia

D.D. Sanyoto
Medical Anatomy Department, Faculty of Medicine, Lambung Mangkurat University, Banjarmasin, Indonesia

ABSTRACT: Stunting is a failure to achieve optimal growth. Seluang fish (*Rasbora* spp.) contains a high amount of protein and calcium needed for bone growth. The purpose of this study was to provide evidence that seluang fish can be a source of nutrition to prevent stunting. The study was conducted on *Rattus norvegicus* that were divided into three treatment groups: C (control), P1 (rats fed with a low-protein and low-fat meal), and P2 (rats fed with a seluang formula). The treatments were given for 8 weeks. Then, the rats were dissected, and blood and bone samples were obtained to measure bone calcium concentration, bone length, and serum IGF-1 level. Statistical analysis was carried out with ANOVA followed by Tukey's honestly significant difference test at the 95% confidence level. The analysis revealed that there were significant differences in bone length (2.92 vs 2.71 vs 2.94 cm, $p < 0.01$), bone calcium level (1.68 vs 0.84 vs 1.34 mg/g, $p < 0.01$), and serum IGF-1 level (70.37 vs 91.37 vs 112.97 pg/mL, $p = 0.02$) among the C, P1, and P2 groups. In conclusion, the study indicates that seluang fish has the potential to prevent stunting.

Keywords: stunting, IGF-1, bone calcium, seluang fish (*Rasbora* spp.)

1 INTRODUCTION

There are many nutrition problems in Indonesia, including protein and energy deficiency, anemia, Vitamin A Deficiency (VAD), iodine deficiency, and overweight. One outcome of the deficiency of protein and energy is stunting. Stunting is a failure to achieve optimal growth, which is measured by the height-for-age index. The Indonesian Basic Health Research conducted in 2013 (Riskesdas 2013) reported that the prevalence of stunting in infants was 37.2% (an increase from 35.6% in 2010). The prevalence of severe stunting only slightly decreased from 18.5% in 2010 to 18.0% in 2013, while that of stunting increased to 19.2%. In addition, among the 33 provinces in Indonesia, 20 provinces had prevalence rates of stunting above the national average. South Kalimantan was ranked as the fifth highest province in the prevalence rate of stunting (about 40%) (Kemenkes RI, 2013). These data showed that the problem of malnutrition, especially stunting, in South Kalimantan has not yet been solved.

Stunting is a chronic malnutrition or growth failure and used as an indicator of long-term malnutrition among children. Children who have moderately poor or poor nutritional status and short or very short stature are at risk of having a lower Intelligence Quotient (IQ) by 10–15 points (Kementerian Bappenas, 2011). Many factors are associated with stunting in children, such as lack of energy and protein, chronic diseases, feeding mistake, and poverty. The prevalence of stunting increases with age. The increase occurs in the first two years of life when the growth of children reflects the standards of nutrition and health.

One of the factors that directly determines the nutritional status of infants and toddlers is inadequate nutrition that does not meet the amount and composition of the required nutrients, especially in the first 1,000 days of life. A crucial period in the development of metabolism and cognition in infants occurs during the first 1,000 days of life. This period can be divided into three phases: pregnancy (9 months), exclusive breastfeeding (6 months), and complementary feeding (18 months). Good nutrition during pre-pregnancy, pregnancy, lactation, and complementary feeding is the main factor that determines the nutritional status of infants, which can be anticipated early by using resources and local wisdom.

South Kalimantan is one of the provinces in Indonesia that had an increase in the number of short toddlers (stunting) in 2013 (43%) compared with 2010. This rate remained above the target of the Medium-Term National Development Plan (RPJMN) in 2014 (32%). National Development Planning Agency (Badan Perencanaan Pembangunan Nasional, Bappenas) found that South Kalimantan is included in stratum 3 group that had the prevalence rate of stunting children under five years old of >32% and the proportion of the population with food insecurity of ≤14.47%, while 39.3% of the population had energy consumption rate below the minimum and 28.0% had protein consumption rate below the minimum (Kemenkes RI, 2010). It is an irony, because South Kalimantan has abundant food resources. One source of food in South Kalimantan is the freshwater (rivers) that contains various types of fish. According to the data from the Department of Fisheries and Marine in South Kalimantan in 2010 (Dinas Perikanan dan Kelautan, 2010), the level of fish consumption by South Kalimantan population was 36.84 kg/person/year. This estimate is higher than the national fish consumption (33.89 kg/person/year), but it is still below that of Malaysia (55.4 kg/person/year) and Singapore (37.9 kg/person/year).

Fish is a good source of protein and calcium, especially fish with edible flesh and bones. Seluang fish (*Rasbora* spp.) is a river fish that is widely consumed by the population in South Kalimantan and included as a type of fish endemic to Borneo and Sumatra. Nutrient content per 100 g of type of fish amounts to 361 kcal, 10 g protein, 3.2 g fat, 80 mg calcium, 224 mg phosphorus, and 4.7 mg iron (Komunikasi edukasi dan Jaringan Usaha, 2013). However, the contents of essential amino acids and essential fatty acids are still unknown. The content of these nutrients may differ between the regions of the origin of the fish. Therefore, we conducted a research on the potential of seluang fish with an effort to address the nutritional problems prevailing in Indonesia. This study aimed to determine whether seluang fish in South Kalimantan can be a source of nutrition to prevent stunting by using a rat model (*Rattus norvegicus*).

2 MATERIALS AND METHODS

This study used female white rats (*Rattus norvegicus*) as models. This study was approved by the Animal Care and Experimentation Committee (Ethical Committee), Faculty oh Medicine Lambung Mangkurat University).

White rats were maintained for 1 week before treatment, to provide an equal physical and psychological condition. During maintenance, white rats were given distilled water and the same food as needed (*ad libitum*).

Fresh seluang fish (*Rasbora* spp.) was minced and made into raw fish porridge. The porridge was steamed with hot steam for 1 hour, and then dried in an oven until the moisture content reached 8%. Seluang fish was grinded again to break down the clots or bone particles. After it became dry grain, the seluang fish meal was made into pellets and used as rat feed.

Rats were classified into three treatment groups: Control (C) group, where mice were fed with a standard diet; treatment group 1 (P1), where mice were fed with a low-protein and low-fat meal; and treatment group 2 (P2), where mice were fed with a seluang fish meal. Feeding was started at 4 weeks of age and given *ad libitum* until 12 weeks. The P1 group was fed with pellets of *nasi karak* (dry spoiled rice), equivalent to a 4% low-protein feed (Illiandri, 2010). The P2 group was fed with the seluang fish meal, containing 25% standard feed mixed with 75% fish meal, and then and made into pellets.

Blood samples were obtained from the heart, and then the samples were centrifuged at 3000 rpm for 10 minutes to obtain serum. Serum was re-centrifuged at 6000 rpm for 10 minutes to obtain the supernatant. The supernatant was then used to measure the levels of IGF-1. Serum IGF-1 levels were measured using ELISA, according to the manufacturer's instructions. Absorbance was measured using an ELISA reader at a wavelength of 450 nm.

Long bones (femurs) were also obtained to determine calcium levels in the bones. Left femur bones were crushed and homogenized, and calcium levels were measured by using the titrimetric method. Scales and caliper were used to measure the right femur bone length. It was measured from the femoral head to the distal of the femur.

The normal distribution of the data was tested using the Shapiro–Wilk normality test. Despite data transformation, some data were still not normally distributed and not homogeneous. For non-normally distributed data, the differences between the groups were analyzed by the Kruskal–Wallis test, followed by the Mann–Whitney test. For normally distributed data, one-way ANOVA was used to assess the differences, followed by Tukey's Honestly Significant Difference (HSD) test. All analyses were conducted at the significance level of 5%.

3 RESULTS

After 8 weeks of treatment, the rats were killed, and the blood and bone samples were collected. The results of the measurement of body weight, long bones, bone calcium levels, and serum IGF-1 levels are respectively shown in Figures 1 to 4.

As shown in Figure 1, there is a difference in the average body weight among the three groups, in which the average weight gain in the P1 group was smaller than that in the other groups. The Kruskal–Wallis test showed a significant difference in body weight among the groups (p < 0.01). Subsequently, the Mann–Whitney test showed that the body weight of the P1 group was significantly different from that of the C group (p = 0.01) and the P2 group (p = 0.02), but no significant difference

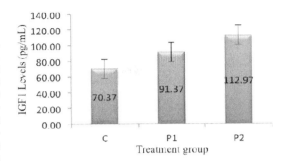

Figure 4. Average IGF-1 levels after 8 weeks of treatment (p < 0.05). C = control group, P1 = low-protein and low-fat diet, P2 = seluang formula diet.

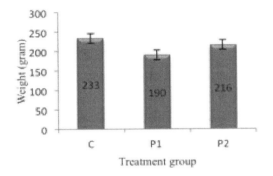

Figure 1. Average body weight after 8 weeks of treatment (p < 0.05). C = control group, P1 = low-protein and low-fat diet, P2 = seluang formula diet.

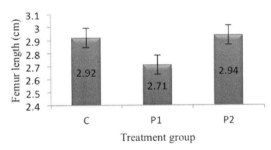

Figure 2. Average femur length after 8 weeks of treatment (p < 0.05). C = control group, P1 = low-protein and low-fat diet, P2 = seluang formula diet.

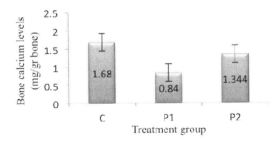

Figure 3. Average bone calcium levels after 8 weeks of treatment (p < 0.05). C = control group, P1 = low-protein and low-fat diet, P2 = seluang formula diet.

was found between the C group and the P2 group (p = 0.42). These results indicate that the seluang formula has the same amount of energy as the standard feed. Thus, seluang feeding did not lead to weight loss. The low-protein and low-fat diet contained less energy than the standard feed.

The femur length of rats in each treatment group is shown in Figure 2. It shows that the bone length in the P1 group is shorter than that in the other groups. The average femur length in the P2 group is not significantly different from that in the C group. The ANOVA test showed that there were significant differences in the average femur lengths between the treatment groups (p < 0.01). A *post hoc* analysis with Tukey's HSD test revealed that there were significant differences between the P1 group and the C (p = 0.02) and P2 (p = 0.01) groups. The average femur length of rats in the P2 group was not significantly different from that in the C group (p = 0.96). This indicated that the P1 group had nutrient deficiency, especially calcium for bone growth, whereas the P2 group had a sufficient intake of nutrients, especially calcium, so that the average bone length in the P2 group was not significantly different from that in the C group.

Bone growth depends on nutritional factors. Intake of calcium and protein will lead to denser and stronger bones. In this study, the average bone calcium levels after administration of treatment for 8 weeks are shown in Figure 3. The average bone calcium levels in the P1 group were lower than that in the other groups. The Kruskal–Wallis test showed that there were significant differences between the treatment groups (p < 0.01). The Mann–Whitney test showed that the bone calcium levels between the C, P1, and P2 groups were significantly different (p < 0.05). The P1 group had the lowest bone calcium levels. This is due to the low calcium content present in the feed. Meanwhile, the P2 group had bone calcium levels lower than those fed with the standard feed. This is because the standard feed consisted of various mixtures of feed materials, while seluang feed content was purely derived from seluang fish.

IGF-1 is a growth factor that has a similar structure to that of insulin. The average level of IGF-1 measured in each group is shown in Figure 4.

The ANOVA test showed significant differences among the treatment groups (p = 0.02). A subsequent analysis with Tukey's HSD test showed that the IGF-1 level in the C group was not significantly different from that in the P1 group (p = 0.30), but was significantly different from that of the P2 group (p = 0.01). Meanwhile, the IGF-1 level in the P1 group was not significantly different from that in the P2 group (p = 0.29). Nevertheless, there was a tendency that the IGF-1 level in the P2 group was the highest among the three groups. This suggested that the intake of nutrients from the seluang meal may increase IGF-1 synthesis, which plays a role in bone growth.

Table 1. Nutrients in seluang fish from South Kalimantan, per 100 grams.

Nutrients	Content (%, w/w)	Nutrients	Content (%, w/w)
Ca	1.6	Histidine	1.82
Fe	19.9 ppm	Arginine	3.05
P	1.67	Threonine	2.15
Zn	122.7 ppm	Alanine	2.83
	11.77	Proline	1.92
Linoleic acid	7.33	Valine	2.41
DHA	1.04	Methionine	0.48
Aspartate	3.71	Isoleucine	2.07
Glutamate	4.98	Leucine	3.62
Serine	1.94	Phenylalanin	2.37
Glycine	3.06	Lysine	4.81
Histidine	1.82	Cysteine	0.32
Tyrosine	1.62		

4 DISCUSSION

Stunting is a failure to achieve optimal growth, which is measured by the height-for-age index. The most effective intervention for stunting is a combination of handling the infection and nutritional programs. Multisectoral cooperation is also needed, namely agriculture sector for food procurement, health sector by giving supplements to pregnant women and children up to the age of 5 years, accompanied by the evaluation of the physical growth of children, and education sector by providing information to the communities. All of these can be carried out properly if they are supported by a good infrastructure (Remans, 2011).

We conducted a study to determine the potential of the existing food, seluang fish, as a source of nutrients to prevent stunting in South Kalimantan. Fish is a good source of protein and calcium, especially fish with its whole flesh and bones that are edible. Seluang fish are well known as a river fish and widely consumed by the population in South Kalimantan, which is included as the fish endemic to Borneo and Sumatra. The study by Noor suggested that each 100 g of seluang fish from Barito River contains 18 g protein, 20 g fat, 52 g water, and 10 g ash (Noor, 1993). The specific nutrient content analyzed by Yunanto et al. in seluang fish from South Kalimantan are presented in Table 1 (Yunanto, 2014).

Based on this analysis, it can be observed that the nutrients in seluang fish are complete, including essential fatty acids and essential amino acids. Calcium in seluang fish from South Kalimantan is about 1.8 g/100 g or 180 mg/100 g dried seluang fish. It can be used as a source of calcium for bone growth, especially in childhood. The calcium content is higher than that present in the same fish species in other areas. Calcium is one of the essential nutrients that is needed for various body functions. Calcium functions as a nutrient essential for growth, contributing to the development of optimal motor function. The bones and teeth contribute up to 99% of the total calcium body, and the rest is found in the blood. When blood calcium level decreases, it will be replaced by bone calcium to maintain the calcium level in the blood (Suptijah et al, 2012; Logesh et al, 2012).

This was proven in this study. The bone lengths of the group given seluang fish were longer than that of the group fed with a low-protein and low-fat meal; however, it was not significantly different from the group fed with the standard feed. The bones of seluang fish are edible, so the fish can meet the calcium requirement. Consumption of calcium with vitamin D from sunlight may enhance the growth of the long bones (arms and legs) and optimize the bone health. Bone health is characterized by a larger bone cell size, density, and better bone remodeling (Prentice et al, 2006).

In this study, we also measured calcium levels in the bones. In people with calcium intake deficiency, the bone density is reduced. This occurs through the regulation of blood calcium levels. When calcium intake is low, the blood calcium level decreases, and there is an immediate response by the pituitary by stimulating the parathyroid glands to produce Parathyroid Hormone (PTH). Parathyroid hormone stimulates the formation of cytokines, namely Interleukin-1 (IL-1), Interleukin-6 (IL-6), and Tumor Necrosis Factor (TNF), in the bones. Cytokines activate osteoclasts to stimulate the absorption of calcium in the bones and release into the blood. Their absorption of calcium from the bones will lead to a decrease in bone density, a condition called osteoporosis (Siki, 2009).

This study showed that the group with low-protein and low-fat intake had the lowest bone calcium content, while there was not much difference in bone calcium content between the group fed the standard feed and the group fed with the seluang meal. Bone density is determined by a dynamic balance between bone formation and resorption processes. When the linear growth and the highest volume of bone mass have been reached, the process of remodeling aims to maintain bone mass (Ganong, 2008; Kurniawan, 2012). Bone tissue growth and development are influenced by genetic factors, nutrition, and hormonal functions that affect the rate of bone growth, shape, and size (Ganong, 2008).

During the growth period, there is a specialized separation area at the end of each long bone (epiphyseal) from the shaft of the bone by a plate of cartilage that actively proliferates, the epiphyseal plate. With the new bone placed at the ends of the bone's shaft by these plates, the length of the bone increases. The width of the epiphyseal plates equals to the speed of bone growth. The widening of the bone is influenced by a number of hormones, but most prominently by the pituitary growth hormone and Insulin-like Growth Factor 1 (IGF-1). Bone growth alignment may occur during the epiphyseal bone separation from the shaft, but it is stalled after the epiphyseal growth fusion with the shaft (epiphyseal closure). Epiphyseal growth plate closure begins in a regular order, and the final epiphyseal closure occurs after puberty. The process of bone development begins with the formation of bone, which is an increase in the number of the basic cell substance. At the same time, the size of the cell increases, as a polyhedral shape, and then they are interconnected through a number of processes of adjacent cells. At this stage, these cells are known as osteoblasts. They will prepare the surface layer of the bone. Bones become thicker by the addition of a matrix layer produced by osteoblast activity (Ganong, 2008).

In this study, we also measured the IGF-1 levels in each treatment group. IGF-1 levels in the group fed with seluang fish were higher than that in the other groups. This suggested that the protein in seluang fish is able to increase the synthesis of IGF-1 that plays a role in the bone formation and growth. A study conducted by Wan Nazaimon et al. (1997) reported that children who were moderately or severely malnourished had lower IGF-1 levels compared with those administered with normal nutrition. The study also found a correlation of IGF-1 levels with the body height and weight in the period before puberty. Malnutrition is known to cause interference rearrangement on the axis of GH/IGF-1, causing an increase in GH levels and a decrease in IGF-1 levels. The study reported a positive and a very significant correlation between the IGF-1 level and its receptor IGFBP3, indicating that both IGF-1 and IGFBP3 simultaneously affect malnutrition. In addition, it has been reported that in terms of malnutrition and growth failure, there was a relationship between the increasing age and IGF-1 levels (Wan Nazaimon et al, 1997).

IGF-1 is one of the peptide hormones or growth hormones that is synthesized in the liver and other tissues, locally acting as a paracrine or autocrine hormone (Puche & Castilla-Cortazar, 2012). These hormones affect the growth and differentiation of cells, including bone cells. Nutritional status affects the serum concentration of IGF-1. The lack of energy and protein causes a resistance to growth hormone. This condition is associated with the failure of a growth hormone signaling receptor, which decreases the synthesis of IGF-1 in the liver. Moreover, the growth hormone and IGF-1 are anabolic hormones, while malnutrition causes catabolism. In a state of catabolism, IGF-1 concentration is low, which will inhibit the growth hormone (Roith, 1997).

During bone growth, IGF-1 is required for osteoblast maturation and function. Through the PI3K pathway, IGF-1 reduces the apoptosis of osteoblasts and triggers osteoblastogenesis through the stabilization of B-catenin, increasing Wnt activity. This effect is associated with complete mitogenic activation, causing an increase in the number of osteoblasts, and increases osteoblastic function and bone formation. IGF-1 induces the synthesis of Receptor Activator of Nuclear Factor κ B Ligand (RANK-L) and improves the function of osteoclasts. IGF-1 also induces the expression of Vascular Endothelial Growth Factor (VEGF) in skeletal cells, and VEGF angiogenesis processes integrate with endochondral formation of bone and osteoblastic differentiation and function (Ahmed & Farquharson, 2010).

IGF-1 synthesis in the liver and other tissues needs some essential amino acids. Seluang fish has complete essential amino acids and thus can meet the requirements of essential amino acids, which is about 12% of the total energy needs. This was evident in the group of rats fed with a low-protein and low-fat diet that had lower levels of IGF-1, short long bones, and low calcium levels. Meanwhile, in the group fed with seluang fish, there were high IGF-1 levels, long bones, and high calcium levels. Thus, it is apparent that there is a correlation between the levels of IGF-1, long bones, and calcium levels.

Physical growth disturbance is associated with the bone health; therefore, the nutrition provided should include nutrients for bones. If bone health is not optimal, there is a risk of suffering from stunted growth, among other things. Nutrients consumed should be started from the intrauterine period, so pregnant women should consume vitamin D, protein, fat, Ca, P, Mg, K, vitamin C, Cu, Zn, and folic acid. After birth, nutrition through breastfeeding and complementary micronutrients after receiving solid foods should be provided. These nutrients can increase the length of the bone, which is reflected in the body height and bone health, and can reduce the risk of fracture. The function of these nutrients is affected by the amount consumed, absorption, and excretion by the body (Prentice et al, 2013). Growth and body size in childhood are correlated with bone mass in the late adolescence phase. Furthermore, the duration of exclusive breastfeeding and bone turnover in the first 6 months is positively correlated with the mass of the spine at the age of 17 years (Molgaard et al, 2011). Thus, the entire community, especially pregnant women, should be encouraged to eat fish. Seluang fish has been shown to have the potential to prevent the incidence of stunting, so it can be used as a functional food ingredient from South Kalimantan.

5 CONCLUSION

Rats fed with seluang fish had significantly longer femur lengths compared with those fed with a low-protein and low-fat diet. Bone calcium levels in the group fed with seluang fish were higher than those in the group fed with a low-protein and low-fat diet, but were lower than those in the group fed with the standard feed. The group fed with seluang fish had higher IGF-1 levels than the group fed with the standard feed and the group fed with a low-protein and low-fat diet. Thus, it indicates that seluang fish has the potential to prevent stunting.

Future research on the potential of seluang fish is necessary. The health sector can collaborate with other sectors such as the Ministry of Maritime Affairs and Fisheries and the Ministry of Trade to promote the potential of seluang fish, so that it can be used as a functional food ingredient from South Kalimantan.

ACKNOWLEDGMENT

This work was supported by a fundamental research grant from KEMENRISTEKDIKTI (Indonesian Ministry of Research, Technology and Higher Education) in 2015 and 2016.

REFERENCES

Ahmed SF & Farquharson C. 2010. The effect of GH and IGF1 on linear growth and skeletal development and their modulation by SOCS proteins. *Journal of Endocrinology* 206: 249–259.

Dinas Perikanan dan Kelautan Provinsi Kalimantan Selatan. 2010. Tingkat konsumsi ikan di Kalimantan Selatan. Banjarmasin: Indonesia.

Ganong WF. 2008. Buku ajar fisiologi kedokteran. Edisi 22. Jakarta:EGC.

Illiandri O, Widjoyanto E, Mintaroem K. 2010. Pengaruh suplemen serbuk daun kelor dalam perbaikan fungsi memori tikus dengan diet rendah protein. *Jurnal Kedokteran Brawijaya* 26 (1): 28–31.

Kementerian Bappenas. 2011. Rencana aksi nasional pangan dan gizi 2011–2015. Kementerian Perencanaan Pembangunan Nasional/Badan Perencanaan Pembangunan Nasional (BAPPENAS). Jakarta: Indonesia.

Kemenkes RI. 2010. Riset Kesehatan Dasar 2010. Jakarta: Indonesia.

Kemenkes RI. 2013. Riset Kesehatan Dasar 2013. Jakarta: Indonesia.

Komunikasi Edukasi dan Jaringan Usaha. 2013. Isi kandungan gizi ikan seluang—komposisi nutrisi bahan makanan. Available from Keju.blogspot.com.

Kurniawan LA, Atmomarsono U, Mahfudz LD. 2012. Pengaruh berbagai frekuensi pemberian pangan dan pembatasan pakan terhadap pertumbuhan tulang ayam boiler. *Agromedia* 30(2): 14–22.

Logesh AR, Pravinkumar M, Raffi SM, Kalaiselvam M. 2012. Calcium and phosphorus determination in bones of low value fishes, *Sardinella longiceps* (Valenciennes) and *Trichiurus savala* (Cuvier), from Parangipettai, Southeast Coast of India. *Asian Pac J Trop Dis.* S254–S256.

Molgaard C., Larnkjaer A., Mark BA., Michaelsen KF. 2011. Are early growth and nutrition related to bone health in adolescence? The Copenhagen Cohort Study of infant nutrition and growth. *Am J Clin Nutr* 94(suppl): 1865S-9S.

Noor AM. 1993. Aspek reproduksi ikan Saluang (*Rasbora spp.*) yang tertangkap di perairan Sungai Barito Desa Bantuil Kecamatan Cerbon Kabupaten DATI II Batola. Banjarbaru: Fakultas Perikanan.

Prentice A., Schoenmakers I., Laskey MA, de Bono S., Ginty F., Goldberg GR. 2006. Nutrition in growth and development: nutrition and bone growth and development. *Proc Nutr Soc* 65(4): 348–360.

Prentice AM., Ward KA., Goldberg GR., Jarjou LM., Moore SE., Fulford AJ., and Prentice A. 2013. Critical windows for nutritional interventions against stunting. *Am J Clin Nutr* 97: 911–8.

Puche JE, Castilla-Cortazar I. 2012. Human conditions of Insulin-like Growth Factor-I (IGF-I) deficiency. *Journal Of Translational Medicine* (10) 224: 1–29.

Remans R. 2011. Multisector intervention to accelerate reductions in child stunting: an observational study from 9 sub-Saharan African countries. *Am J Clin Nutr* 94: 1632–42.

Roith DL. 1997. Insulin-like Growth Factors. Seminar of Medicine in the Beth Israel Decones Medical Center. *The New England Journal of Medicine* 336(9): 633–640.

Siki K. 2009. Osteoporosis patogenesis diangnosis dan penanganan terkini. *Jurnal Penyakit* Dalam. 10(2).

Suptijah P, Jacoeb AM, Deviyanti M. 2012. Karakterisasi dan biovailabilitas nanokalsium cangkang udang vannamei (*Litopenaeus vannamel*). *Jurnal Akuatika* 3(1): 63–73.

Wan Nazaimon WM, Rahmah R, Osman A, Khalid BAK, Livesey J. 1997. Effects of childhood malnutrition on Insulin-like Growth Factor-1 (IGF-I) and IGF-1 binding protein-3 level: a Malaysian and New Zealands analysis. *Asia Pasific J Clin Nutr* 6(4): 273–276.

Yunanto A., Sanyoto DD., Triawanti., Syahadatina M., Oktaviyanti IK. 2014. Benefit of seluang fish (*Rasbora spp.*)'s South Kalimantan to the improvement of spatial memory quality. *The 3rd International Symposium on Wetlands Enviromental Management, Banjarmasin 8–9 November 2014*. Indonesia.

Effect of cryoprotectants on sperm vitrification

R. Widyastuti
Laboratory of Animal Reproduction and Artificial Insemination, Department of Animal Production, Animal Husbandry Faculty, Universitas Padjadjaran, Sumedang, West Java, Indonesia

R. Lesmana
Physiology Division, Department Anatomy and Biology Cell, Faculty of Medicine, Universitas Padjadjaran, Sumedang, West Java, Indonesia

A. Boediono
Laboratory of Embryology, Department of Anatomy, Physiology and Pharmacology, Faculty of Veterinary Medicine, Institute of Bogor Agriculture, Jl. Agatis Dramaga Bogor, West Java, Indonesia

S.H. Sumarsono
Physiology, Developmental Biology and Biomedical Science Research Group, School of Life Science and Technology, Bandung Institute of Technology, West Java, Indonesia

ABSTRACT: One of the problems of using high concentrations of cryoprotectants for sperm vitrification is the cytotoxic effect that affects sperm recovery after warming. Therefore, in this study, we determined the recovery rate after vitrification with and without cryoprotectants. Ejaculates with progressive motility and viability above 50% were used as samples. The samples were divided into two groups: (1) samples were mixed with a basic solution (2) samples were mixed with a vitrification solution. Sperm were vitrified by direct plunging into LN2. Sperm motility and viability were observed to evaluate the quality of sperm before and after vitrification. Overall, the sperm samples vitrified with cryoprotectants had a significantly higher proportion of sperm motility (56%) and viability (58.15%) compared with those vitrified without cryoprotectants (35% and 48%, respectively, $p < 0.05$). However, vitrification of human sperm without cryoprotectants could be recommended for routine assisted reproductive technology.

Keywords: sperm vitrification, cryoprotectants, sperm motility, sperm viability

1 INTRODUCTION

Vitrification is known to establish a glass-like solid state during the cooling process. It also has an economic advantage compared with the slow freezing method such as a lack of ice crystal formation due to an increase in the speed of temperature conduction, which provides a significant increase in cooling rates. This rapid cooling process circumvents the ice crystalline formation phase by converting solutions or water into a glass-like amorphous solid (Dinnyes et al., 2007). Vitrification conditions can be achieved by using high concentrations of cryoprotectants. The high concentration of cryoprotectants and extremely rapid rates of cooling are responsible for the formation of the solid state, thereby preventing the formation of intracellular ice crystals. One of the disadvantages of using high concentrations of cryoprotectants for sperm vitrification is the cytotoxic effect (Özkavukcu and Erdemli, 2002), which affects sperm recovery rates after the vitrification process. Therefore, it is particularly detrimental to patients who have low counts of sperm.

The use of cryoprotectants in the vitrification of human sperm requires further study because sperm is gamete cells having an intracellular matrix with high viscosity that may function as an internal cryoprotectant. Moreover, sperm has a compact structure and a small number of cytosols. This structure determines whether the sperm requires a small amount of intracellular cryoprotectants or not. The aim of this study was to observe the recovery rate of human sperm that were vitrified using different methods: (1) with cryoprotectants and (2) without cryoprotectants. Furthermore, this study determined whether the use of toxic intracellular cryoprotectants for sperm vitrification can be avoided and whether sperm vitrification without cryoprotectants can be suitable for use in assisted reproductive technology.

2 MATERIALS AND METHODS

2.1 Sample

Ejaculates were obtained from 20 men as donors by masturbation after 2–7 days of sexual abstinence. The ejaculates were selected after liquefaction for 30 minutes, and those having a concentration of 15 million or more sperm/ml and showing at least 50% progressive sperm motility and viability were used as samples. Semen analysis was performed according to the published guidelines by the World Health Organization (2010). Each ejaculate was divided into three equal parts. Part 1 (P1) was the control group. Part 2 (P2) and part 3 (P3) were diluted with Earle's Balanced Salt Solution (EBSS) medium, and then centrifuged at 600 g for 10 min to remove seminal plasma. After centrifugation, sperm pellets were diluted with 500 μl EBSS medium. The suspended sperm was again diluted (1:1) with a vitrification medium containing (P2) EBSS (without cryoprotectants) and (P3) EBSS + 0.25 M sucrose + 1% Human Albumin Serum (HAS).

2.2 Vitrification and warming

After dilution, the P2 and P3 groups were allowed to equilibrate at room temperature for 10 minutes before vitrification, and then sperm motility and viability were assessed. Afterwards, each specimen was loaded into a 0.25 ml plastic straw using a syringe and then sealed. The straws were vaporized in liquid nitrogen for 5 seconds, and then plunged into liquid nitrogen directly and stored until 24 hours. The straw was taken out from liquid nitrogen, warmed at 37°C water for 5 seconds, and then the tip of the straw was cut, and the sample was put into a microtube. Sperm motility and viability were evaluated.

2.3 Evaluation of sperm motility and viability

Sperm motility was evaluated immediately after liquefaction, dilution with the vitrification medium, and after warming the samples. Here, we used two different categories of sperm motility, i.e., 'a' category for motile sperm and 'b' category for immotile sperm. Sperm viability was observed under a microscope using eosin–nigrosin staining.

2.4 Statistical analysis

Statistical analysis was performed using Minitube version 14. Data were analyzed using ANOVA and Tukey's test with a significance level of $p < 0.05$.

Table 1. Sperm motility during the vitrification process.

Stage of vitrification	Sperm motility (%)		
	P1	P2	P3
Equilibration	100.0 ± 0.0%[a]	88.50 ± 6.30%[a]	94.20 ± 3.40%[a]
Warming	100.0 ± 0.0%[b]	56.90 ± 4.38%[c]	35.00 ± 10.30%[d]

*Values for the same volume with different letters are significantly different ($P < 0.05$).
**P1: control, P2: EBSS (without cryoprotectants), P3: EBSS + 0.25 M sucrose + 1% HAS (with cryoprotectants).

Table 2. Sperm viability during the vitrification process.

Stage of vitrification	Sperm viability (%)		
	P1	P2	P3
Equilibration	100.0 ± 0.0%[a]	92.1 ± 5.50%[a]	95.90 ± 2.00%[a]
Warming	100.0 ± 0.0%[a]	58.15 ± 4.70%[b]	48.06 ± 3.20%[c]

*Values for the same volume with different letters are significantly different ($P < 0.05$).
**P1: control, P2: EBSS (without cryoprotectants), P3: EBSS + 0.25 M sucrose + 1% HAS (with cryoprotectants).

3 RESULTS

The results of sperm motility and viability during the vitrification process are summarized in Table 1 and Table 2. Sperm quality after vitrification was decreased in both methods compared with the control group. Overall, the sperm samples vitrified with cryoprotectants had a significantly higher proportion of sperm motility (56%) and sperm viability (58.15%) compared with those vitrified without cryoprotectants (35% and 48%, respectively, $p < 0.05$).

4 DISCUSSION

There are two types of cryoprotectant: intracellular cryoprotectant that will cross the cell membrane to buffer the intracellular salt, and extracellular cryoprotectant that has an important role in the cell dehydration process. Human sperm contains a large amount of protein, sugar, and other components that make the intracellular matrix highly viscous. As a result, it can be speculated that we can achieve intracellular vitrification for human sperm (Best, 2015, Isachenko et al., 2004). The high protein content in the cells leads to high viscosity, so it is highly sufficient to achieve intracellular vitrification. Thus, high concentrations of cryoprotectants are necessary for extracellular vitrification than intracellular vitrification.

In our study, vitrification of sperm using 0.25 M sucrose and 1% HAS as cryoprotectants showed higher recovery rates of sperm than that without cryoprotectants because sucrose and HAS act as extracellular cryoprotectants that stabilize sperm membrane during the vitrification process. Extracellular cryoprotectants have a size that is too large to diffuse into cells, but they help water vitrification and devitrification, inhibiting the extracellular space. Sucrose is a type of sugar that cannot diffuse into the plasma membrane. Therefore, sucrose will produce an osmotic pressure that induces the dehydration process and reduces the formation of intracellular ice crystals. Sucrose also interacts with phospholipids in the plasma membrane, so that sperm membranes will reorganize the survival rate during freezing. High viscosity will increase the glass transition temperature in the cytosol of sperm.

Based on Table 1 and Table 2, human sperm vitrification without cryoprotectants can still be applied, because human sperm contains a large number of proteins, sugars, and other components that make the intracellular matrix highly viscous and to have compartments that enable the function as natural cryoprotectants (Isachenko et al., 2004b). These results are supported by (Isachenko et al., 2004a), indicating that frozen sperm motility without the use of cryoprotectants in the vitrification method showed a decrease of about 40% when compared with conditions before freezing. A previous study has reported a novel vitrification method for a single spermatozoon using the Cryotop method. The vitrification and warming techniques were simple, and the recovery rate of sperm after warming was relatively high (Endo et al., 2011). The size of the sperm also plays a role in the success of vitrification without cryoprotectants. The sperm is smaller than the embryo, so it is less likely that intracellular ice nucleation is followed by the growth of crystals during the freezing stage, or recrystallization during the warming stage (Isachenko et al., 2003).

5 CONCLUSION

Vitrification of human sperm can be achieved without intracellular cryoprotectants. Moreover, vitrification by directly plunging into liquid nitrogen without cryoprotectants was effective and could be recommended for routine assisted reproductive technology.

REFERENCES

Best, B. P. 2015. Cryoprotectant toxicity: facts, issues, and questions. *Rejuvenation Research* 18: 422–436.

Dinnyes, A. Liu, J. & Nedambale, T. L. 2007. Novel gamete storage. *Reprod Fertil Dev* 19: 719–31.

Endo, Y. Fujii, Y. Shintani, K. Seo, M. Motoyama, H. & Funahashi, H. 2011. Single spermatozoon freezing using cryotop. *Journal of Mammalian Ova Research* 28: 47–52.

Gao, D.Y. Liu, J. Liu, C. Mcgann, L. E. Watson, P. F. Kleinhans, F. W. Mazur, P. Critser, E. E. & Critser, J. K. 1995. Andrology: Prevention of osmotic injury to human spermatozoa during addition and removal of glycerol. *Human Reproduction* 10: 1109–1122.

Isachenko, E. Isachenco, V. Katkov, I.I. Dessole, S & Nawroth, F. 2003. Vitrification of mammalian spermatozoa in the absence of cryoprotectants: from past practical difficulties to present success. *Reproductive BioMedicine Online* 6: 191–200.

Isachenko, E. Isachenko, V. Katkov, I. I. Rahimi, G. Scondorf, T. Mallmann, P. Dessole S & Nawroth, F. 2004a. DNA integrity and motility of human spermatozoa after standard slow freezing versus cryoprotectant-free vitrification. *Human Reproduction* 19: 932–939.

Isachenko, V. Isachenko, E. Katkov, I.I. Montag, M. Dessole, S. Nawroth, F & Vander Ven, H. 2004b. Cryoprotectant-free cryopreservation of human spermatozoa by vitrification and freezing in vapor: effect on motility, DNA integrity, and fertilization ability. *Biology of Reproduction* 71: 1167–1173.

Moskovtsev, S. I. Lulat, A. G. M. & Librach, C. L. 2012. Cryopreservation of Human Spermatozoa by Vitrification vs. Slow Freezing: Canadian Experience. In: Katkov, I. I. (ed.) *Current Frontiers in Cryobiology*. InTech.

Organization, W. H. O. 2010. WHO laboratory manual for the examination and processing of human semen—See more at: http://apps.who.int/iris/handle/10665/44261?mode=simple—sthash.vxXOfoMi.dpuf. 5th ed.: Geneva: World Health Organization.

Ozkavukcu, S & Erdemli, E. 2002. Cryopreservation: basic knowledge and biophysical effects. *Journal of Ankara Medical School* 24: 187–196.

Mucoprotective effect of *Trigona* propolis against hemorrhagic lesions induced by ethanol 99.5% in the rat's stomach

V. Yunivita
Department of Phamarcology and Therapy, Faculty of Medicine, Universitas Padjadjaran, Bandung, Indonesia

C.D. Nagarajan
Medical Student of Faculty of Medicine, Universitas Padjadjaran, Bandung, Indonesia

ABSTRACT: *Trigona* propolis is known to contain a wide range of substances that are beneficial to humans. However, there are no data on the effect of *Trigona* propolis on hemorrhagic gastritis. This study aims to understand the mucoprotective effect of *Trigona* propolis against hemorrhagic lesions in the rat's stomach. The study used 21 male rats that were divided into three experimental groups. Group 1 was given propylene glycol, group 2 was given *Trigona* propolis (300 mg/kg BW), and group 3 was given ranitidine (50 mg/kg BW). All rats were also given ethanol 99.5% (1 mL) one hour later. On the 3rd day, the rats were dissected, the stomach was incised, and the characteristics of hemorrhagic lesions were recorded. The average lengths of hemorrhagic lesions in the three groups were 81.86 ± 15.52 mm, 43.29 ± 9.23 mm, and 60.14 ± 12.88 mm, respectively, and the differences were significant between the groups ($p < 0.05$; one-way ANOVA). This study showed that *Trigona* propolis has a mucoprotective effect and therefore can be used as a preventive treatment for hemorrhagic gastritis induced by ethanol 99.5%.

1 INTRODUCTION

Peptic ulcers are sores in the mucous lining of the lower esophagus, stomach, or proximal duodenum that reach the muscularis mucosae layer. Peptic ulcer is defined as an imbalance between acid and pepsin secretion, and the digestive tract is unable to protect itself from damage (Syam, Sadikin, Wanandi, & Rani, 2009). Some factors can increase the risk of developing peptic ulcer resulting from regular consumption of anti-inflammatory drugs (NSAIDs) and infection by *Helicobacter pylori* (Sung, Kuipers, & El-Serag, 2009).

Peptic ulcer disease was considered a rare disease in the 1800s, but its prevalence has increased since the first half of the 20th century (Bertleff & Lange, 2010). Through the years of medical experience, many therapies have been developed for the treatment of peptic ulcers in the hope of lowering mortality rates due to peptic ulcer disease. Generally, the treatments for peptic ulcers aim to neutralize acid in the stomach, block acid production by blocking histamine release or inhibition of proton-pump, protect the mucus lining by forming a barrier between the ulcer and the acidic environment, and increase mucus production (Sung et al., 2009). However, these drugs tend to cause a fair amount of side effects and adverse effects.

One disease that has the potential to develop into peptic ulcers is gastritis. Symptoms of gastritis are less severe than those of peptic ulcers. However, it should not be taken lightly since symptoms that are left untreated can and will develop a form of gastritis known as erosive gastritis or better known as hemorrhagic gastritis. Severe forms of hemorrhagic gastritis can lead to ulcer formation in certain cases (Rugge et al., 2011).

The medical field has always aimed to improve the quality of human life. This is why the testing of new remedies is always welcomed. Natural remedies are a well-known therapy of various diseases among older generations, enabling health professionals to believe that there may be benefits from treatment with natural substances. Concerning peptic ulcer disease, propolis stands out as a potential natural remedy; moreover, since hemorrhagic gastritis is generally treated with same medications that are used in the treatment of peptic ulcers, propolis is relevant even in this aspect (Pillai, Kandaswamy, & Subramanian, 2010). According to the Roman scholar Plinius (23–79 A.D.), propolis can relieve pain and heal sores that appeared impossible for physicians to correct at that time (Bogdanov, 2014).

Propolis is a strong adhesive viscous material that is collected by bees from various plant sources.

In Indonesia, one of the bee species from which propolis is obtained is the *Trigona* species. This species is commonly found in Asia and Africa. *Trigona* is a stingless bee and the smallest in size among other honey-producing bees. Their nests are usually found in tree hollows, wooden tables, holes in the wall, and even on steel supports. *Trigona* is known as a good propolis producer (Kustiawan et al., 2015).

Generally, propolis contains more than 300 components depending on its geographical origin. Each component works together for the health and protection of the beehive. The active components of propolis have also been proven effective in the health of the human body. Until now, the administration of propolis to mice and humans has not shown any side effects (Bogdanov, 2014; Pillai et al., 2010).

Propolis has many beneficial effects, and one of them is its anti-ulcer (stomach, skin, buccal) effect. Other studies have shown that some chemical compounds found in propolis, such as polyphenols, flavonoids and caffeic acid, have the ability to exert an anti-ulcer effect (Bogdanov, 2014). These compounds lead to the esterification of membrane phospholipids that will produce arachidonic acid that forms the enzyme Cyclooxygenase (COX) which stimulates the production of prostaglandins (precisely prostaglandin-E$_2$). This will inhibit the acid-secreting parietal cells present in the stomach, resulting in decreased gastric acid secretion (Pillai et al., 2010). A decrease in gastric acid secretion can be the factor for the prevention of hemorrhagic gastritis and, in the long run, avoiding the occurrence of peptic ulcers. Therefore, this study aims to investigate whether propolis has a protective effect against the occurrence of hemorrhagic gastritis in order to be used as a natural remedy for peptic ulcer.

2 MATERIALS AND METHODS

2.1 Drugs and chemicals

Trigona propolis extract (300 mg/kg BW) was obtained from the Laboratory of Food Engineering, Faculty of Food Industry Technology, Universitas Padjadjaran. Propylene glycol used in the extraction process of propolis as well as ethanol 99.5% were obtained from PT. BRATACO, Bandung and Ranitidine was from PT. Nicholas Laboratories Indonesia, Jakarta, Indonesia.

2.2 Experimental animals

A total of 21 healthy, male rats of the Wistar strain (weighing 150–200 g) were used in this study. The rats were obtained from Pusat Antar Universitas Institut Teknologi Bandung (PAU ITB) through the Animal Laboratory of Pharmacology and Therapy, Faculty of Medicine, Padjadjaran University, Bandung. The study was approved by the Independent Ethics Committee of Faculty of Medicine of Universitas Padjadjaran (Bandung, Indonesia). The rats were randomly divided into three experimental groups, with seven rats per group. The animals were housed in cages, and had *ad libitum* access to a standard pellet diet and water.

2.3 Study of the effect of Trigona propolis using an ethanol-induced ulcer model

The rats were fasted for 12 hours before the experiment. Group 1 was given 1 mL propylene glycol, group 2 was given *Trigona* propolis extract (300 mg/kg BW), and group 3 was given ranitidine (50 mg/kg BW). At 1 hour later, the rats were given 1 mL ethanol 99.5% by orogastric intubation to induce gastric ulcers, according to the method described previously by Pillai et al. (2010) with slight modification. All pretreatments were administered orally. Then, the rats were given food and water as usual.

The above treatment procedure was carried out for 3 days. On the 3rd day, one hour after the administration of ethanol, all the rats were killed by cervical dislocation. A midline incision was made across the abdomen of the rat and the stomach was moved to the exterior of the body. The stomach was then opened by an incision along the greater curvature and gently rinsed with water, pinned flat on a wood plate and fixed in 10% formalin for 24 hours. The stomach was examined under a microscope and all hemorrhagic lesions were counted and measured (in mm) with a ruler (Dugani, Auzzi, Naas, & Megwez, 2008).

2.4 Determination of hemorrhagic lesions

Ulcers found in the gastric mucosa appeared as elongated bands of hemorrhagic lesions parallel to the long axis of the stomach. The stomach was examined under a microscope, and the number of hemorrhagic lesions was counted and the length of the hemorrhagic line (in mm) was measured with a ruler (Dugani et al., 2008).

2.5 Statistical analysis

All values are reported as means ± SD. The statistical significance of differences between groups was assessed using one-way ANOVA. A probability value of $p < 0.05$ was considered to be statistically significant.

3 RESULTS AND DISCUSSION

Based on the outcome of the study, the mean value was calculated for the number of hemorrhagic lesions and the length of the hemorrhagic line induced by ethanol in the rat's stomach. The data were evaluated to ensure normal distribution and homogeneity of the variance. Since the data showed a normal distribution (p > 0.05) and a homogeneous variance (number of hemorrhagic lesions: p = 0.557, length of the hemorrhagic line: p = 0.482), the ANOVA test was performed to determine the significance of data between the groups. Then, a Bonferroni *post hoc* test was done to compare the effect of each group, assuming homogeneity of the variance.

The results obtained for the number of hemorrhagic lesions and the length of the hemorrhagic line are summarized in Table 1.

From Table 1, it can be seen that group 2 administered with *Trigona* propolis had the smallest average number of hemorrhagic lesions (4 ± 1.160, p = 0.062), with no significant difference between the three groups.

The length of the hemorrhagic line in the group that received *Trigona* propolis revealed the lowest value (43.29 ± 9.23, p < 0.001), as shown in Figure 1.

Table 1. Comparison of stomach hemorrhagic lesions between the groups.

	Propylene glycol + ethanol 99.5% (n = 7)	*Trigona* propolis 300 mg/kg BW + ethanol 99.5% (n = 7)	Ranitidine 50 mg/kg BW + ethanol 99.5% (n = 7)	p
Number of lesions	5.57 ± 1.13	4 ± 1.16	5.29 ± 1.38	0.062
Length of the line (mm)	81.86 ± 15.52	43.29 ± 9.23	60.14 ± 12.88	<0.05

One-way ANOVA test.

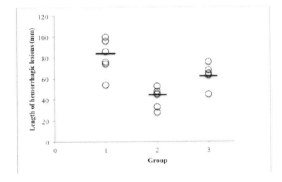

Figure 1. Distribution of the length of the hemorrhagic line.

ANOVA test revealed a significant difference between the groups. The Bonferroni *post hoc* test at a 95% confidence interval revealed that the length of the hemorrhagic line was significantly different between the groups that received *Trigona* propolis or ranitidine and propylene glycol (p < 0.05). The length of the hemorrhagic line was not significantly different between the groups that received *Trigona* propolis and ranitidine. Destruction of mucosal lining occurs as a result of an imbalance between aggressive factors such as acid and pepsin production, and defense mechanisms, like mucus and bicarbonate production, blood flow, and mucosal turnover. Ethanol 99.5%, a form of alcohol, causes irritation in the stomach mucosa that can lead to hemorrhagic gastritis. Gastric damage due to ethanol can be associated with stasis of blood flow that results in hemorrhage, one of the aspects of tissue injury (Pillai et al., 2010). Studies have suggested that ethanol had a direct toxic effect by making disruption of the vascular endothelium resulting in increased vascular permeability, reducing secretion of bicarbonates, and production of the mucus (Abdulla, AL-Bayaty, Younis, & Abu Hassan, 2010). As demonstrated in this study, ethanol was used to induce mucosal damage. The results showed that the control group had the highest average number of hemorrhagic lesions as well as the highest average length of the hemorrhagic line. This indicates that ethanol 99.5% does cause irritation in the gastric mucous.

Ranitidine is a commonly used gastrointestinal drug that is used for the prevention of peptic ulcer (Yeomans, Svedberg, & Naesdal, 2006). Therefore, ranitidine was used in this study to compare the effect of propolis with this drug. The result from this study revealed that the average length of the hemorrhagic line in this group was significantly lower than that of the control group (group 1). This indicates that ranitidine has a preventive effect, in which it reduces the length of the hemorrhagic line.

Studies of propolis conducted in various countries such as Brazil and India have suggested that propolis has a mucoprotective effect (Massignani et al., 2009; Pillai et al., 2010). To test propolis from Indonesia, the rats in group 2 were given *Trigona* propolis 300 mg/kg BW followed by ethanol 99.5% one hour later. The results showed a significant protective effect towards ethanol-induced mucosal irritation, in which the average number of hemorrhagic lesions and the length of the hemorrhagic line were 4 ± 1.16 and 43.29 ± 9.23, respectively. This is consistent with the result of the study conducted on Indian propolis (Pillai et al., 2010). Compared with the group fed with propylene glycol, it can be observed that *Trigona* propolis has an anti-ulcerogenic effect related to its cytoprotective activity, since it significantly reduced ethanol-induced ulcers. This effect was similar to the effect of ranitidine. From this study, it is found that *Trigona* propolis also has a comparatively similar effect to ranitidine in preventing hemorrhagic gastritis. Many studies have suggested that phytochemicals such as flavonoids and phenolic compounds, which are well known for their anti-ulcer activity and other antioxidant compounds, could be active in this experimental model to produce an anti-ulcerogenic effect (Pillai et al., 2010). Therefore, the observed ulcer preventive and ulcer curative effect of *Trigona* propolis may be partially due to its relative antioxidant activity.

A limitation of this study is the relatively small number of animals in each experimental group.

4 CONCLUSION

This study proves that *Trigona* propolis, a type of propolis in Indonesia, has a mucoprotective effect against hemorrhagic lesions induced by ethanol 99.5% in the rat's stomach, and it could be a good therapeutic agent for the development of a new phytotherapeutic medicine for the treatment of gastric ulcer. Further studies are needed to explore the mucoprotective effects of chronic administration of *Trigona* propolis in the ethanol model of stomach hemorrhagic lesions and in other models, such as indomethacin-induced ulcers, using the ulcerative lesion index and histological evaluation.

REFERENCES

Abdulla, M. A., AL-Bayaty, F. H., Younis, L. T., & Abu Hassan, M. I. (2010). Anti-ulcer activity of Centella asiatica leaf extract against ethanol-induced gastric mucosal injury in rats. *Journal of Medicinal Plants Research*, 4(13), 1253–1259.

Bertleff, M. J. O. E., & Lange, J. F. (2010). Perforated peptic ulcer disease: A review of history and treatment. *Digestive Surgery*.

Bogdanov, S. (2014). Propolis : Composition, Health, Medicine : A Review. *Bee Product Science*, 1–40.

Dugani, A, Auzzi, A, Naas, F., & Megwez, S. (2008). Effects of the oil and mucilage from flaxseed (*linum usitatissimum*) on gastric lesions induced by ethanol in rats. *The Libyan Journal of Medicine*, 3(4), 166–9.

Kustiawan, P. M., Phuwapraisirisan, P., Puthong, S., Palaga, T., Arung, E. T., & Chanchao, C. (2015). Propolis from the stingless bee trigona incisa from East Kalimantan, Indonesia, induces in vitro cytotoxicity and apoptosis in cancer cell lines. *Asian Pacific Journal of Cancer Prevention*, 16(15), 6581–6589.

Massignani, J. J., Lemos, M., Maistro, E. L., Schaphauser, H. P., Jorge, R. F., Sousa, J. P. B., Andrade, S. F. (2009). Antiulcerogenic activity of the essential oil of Baccharis dracunculifolia on different experimental models in rats. *Phytotherapy Research*, 23(10), 1355–1360.

Pillai, I., Kandaswamy, M., & Subramanian, S. (2010). Antiulcerogenic and ulcer healing effects of Indian propolis in experimental rat ulcer models. *Journal of ApiProduct and ApiMedical Science*, 2(1), 21.

Rugge, M., Pennelli, G., Pilozzi, E., Fassan, M., Ingravallo, G., Russo, V. M., & Di Mario, F. (2011). Gastritis: The histology report. *Digestive and Liver Disease*, 43(Suppl. 4).

Sung, J. J. Y., Kuipers, E. J., & El-Serag, H. B. (2009). Systematic review: the global incidence and prevalence of peptic ulcer disease. *Alimentary Pharmacology & Therapeutics*, 29(9), 938–46.

Syam, A. F., Sadikin, M., Wanandi, S. I., & Rani, A. A. (2009). Molecular mechanism on healing process of peptic ulcer. *Acta Medica Indonesiana*, 41(2), 95–8.

Yeomans, N. D., Svedberg, L. E., & Naesdal, J. (2006). Is ranitidine therapy sufficient for healing peptic ulcers associated with non-steroidal anti-inflammatory drug use? *International Journal of Clinical Practice*, 60(11), 1401–1407.

Author index

Abdulah, R. 53
Achmad, S. 61
Achmad, T.H. 19
A'ini, A.W. 33
Akbar, I.B. 41
Anggraeni, N. 1
Arisanti, N. 5
Arya, I.F.D. 5
Avriyanti, E. 105

Bahrudin, M. 33
Boediono, A. 119

Cahyadi, A.I. 13

Dahlan, A. 9
Damara, F.A. 1, 9
Darjan, M. 29
Dewanto, J.B. 9
Dewi, I.M.W. 13
Dhianawaty, D. 1
Dian Irianti, C.C. 1
Diela, B. 13

Elliyanti, A. 19

Fadhil 33
Farenia, R. 41
Fatimah, S.N. 23, 41

Ghozali, M. 109
Goenawan, H. 37
Gunawan, H. 105

Hanna 41
Hardjadinata, I.S. 29
Harianja, F. 9
Hernowo, B.S. 37
Heryaman, H. 9
Hidayat, T. 41
Hindritiani, R. 79
Huda, F. 41
Hutama, S.H. 57

Indah, Y. 41
Iso, T. 61, 109

Juansah, R.D. 93
Juliati 41
Jusuf, A.A. 71

Kameo, S. 61
Karhiwikarta, W. 41
Kartasasmita, C.B. 13
Khaerunnisa, R. 29
Koyama, H. 53, 61
Kurabayashi, M. 61, 109

Lesmana, R. 37, 119
Lestari, D.Y. 33
Lubis, L. 41

Madjid, T.H. 93
Malini, D. 87
Masjhur, J.S. 19
Maskoen, A.M. 19, 23, 87
Masuda, K. 65
Maulinda, S. 79
Muryani, A. 47

Nagarajan, C.D. 123
Najmi, N. 37
Ninda, J. 41
Noormartany, N. 19
Nugraha, G.I. 23, 41, 109
Nurhayati, T. 41

Panigoro, R. 1, 5, 87, 109
Pratiwi, Y.S. 41
Prihananti, D.H. 57
Prisinda, D. 47
Purba, A. 23, 41
Puspitasari, I.M. 53
Putra, A. 57
Putri, M. 61, 109

Qomarilla, N. 93

Rachmayati, S. 41
Rahayu 33
Rahmalita, A. 57
Rakhimullah, A.B. 1
Ray, H.R.D. 65
Razali, R. 71
Redjeki, S. 71
Reniarti, L. 87
Rizky Akbar, M. 41
Robianto, S. 1
Roesmil, K. 23
Rohmawaty, E. 75
Ronny 41
Ruchiatan, K. 79
Ruslami, R. 75, 83
Ruslina, I. 41

Sa'diyah, N.A.C. 57
Safitri, R. 87
Saktiadi, R.S.P. 93
Sanyoto, D.D. 113
Sastramihardja, H.S. 75
Sazali, A. 99
Setiawan 37, 41
Setiawan, D. 99
Setiawati, D.A. 1
Setiawati, E.P. 5
Shahib, M.N. 75
Siswanti, T.Y. 93
Sribudiani, Y. 19, 37
Sudigdoadi, S. 13, 93
Sumantri, N.I. 99
Sumarsono, S.H. 119
Sunjaya, D.K. 41
Sutadipura, N. 9, 61
Sutedja, E. 79, 93
Suwarsa, O. 105
Syamsunarno, M.R.A.A. 1, 53, 61, 87, 109
Sylviana, N. 41

Tarawan, V.M. 37, 41
Tarra, Y. 57

Tessa, P. 41
Triatin, R.D. 1
Triawanti 113

Utami, N.V. 41

Widyastuti, R. 119
Wikayani, T.P. 93
Wilopo, B.A.P. 13

Yamazaki, C. 61
Yunanto, A. 113
Yunivita, V. 123